INTRODUCTORY REMOTE SENSING: PRINCIPLES AND CONCEPTS

Paul J. Gibson

With contributions to the text by Clare H. Power, and Website development by John Keating

ROUTLEDGE
ROUTLEDGE
Taylor & Francis Group

First published 2000
by Routledge
11 New Fetter Lane, London EC4P 4EE

Simultaneously published in the USA and Canada
by Routledge
29 West 35th Street, New York, NY 10001

Routledge is an imprint of the Taylor & Francis Group

© 2000 Paul J. Gibson

Typeset in Garamond by J&L Composition Ltd, Filey, North Yorkshire
Printed and bound in Great Britain by
St Edmundsbury Press,
Bury St Edmunds, Suffolk

British Library Cataloguing in Publication Data
A catalogue record for this book is available from the British Library

Library of Congress Cataloging in Publication Data
A catalog record for this book has been requested

ISBN 0–415–17024–9 (hbk)
ISBN 0–415–19646–9 (pbk)

This project is for Dot and my family

CONTENTS

COLOUR PLATES

The following plates appear between pp.76–77

FIGURES

TABLES

PREFACE
OUTLINE OF *INTRODUCTORY REMOTE SENSING*

Most of us have been introduced to remote sensing at some stage in our lives. At the shallowest level, this may only involve seeing a satellite image used as a 'pretty picture' adorning the frontispiece of a book or at a deeper level involving imagery of the planets of our Solar System as imaged by spacecraft such as Voyager. However, in seeking to obtain further information (how is the imagery transmitted? what dictates the observed colours? which is the optimum system and so on), one can easily come away with the view that remote sensing is a complex subject composed of such a range of components that gaining an in-depth knowledge and understanding of it is impossible. *Introductory Remote Sensing* seeks to provide a greater understanding of the various aspects of remote sensing by considering this field of science in two volumes. *Introductory Remote Sensing: Principles and Concepts* is the basic text and an outline of what is covered in this book is shown in Figure P.1. This book seeks to address four main questions.

What is remote sensing?
What principles govern remote sensing?
How are remote sensing data obtained?
What are the applications of remote sensing?

This book also includes a dedicated WWW site that has a number of sections. The site can be accessed by Netscape version 3 (or higher) or Internet Explorer version 4 (or higher). This site contains additional remote sensing images encompassing the fields of meteorology; geology; vegetation studies; urban studies; oceanography; and environmental applica-

tions. An important component of remote sensing is the ability to interpret images. Consequently, a number of images are provided on the WWW site which may be used either in a classroom discussion context or a means of determining your competence in image interpretation. Allied to this is a section containing questions (and answers) on remote sensing theory, covering all aspects of this volume. Other sections within the WWW site include a glossary, sources of remote sensing information; a list of data contributors to the books and a preview of the companion volume, *Introductory Remote Sensing: Digital Image Processing and Applications*.

This book has a number of main sections (Figure P.2). Section 1 explains how the digital data may be processed in order to maximise the information output. This section is divided into pre-processing the data (Chapter 2) and enhancing the data (Chapter 3). A greater in-depth discussion of environmental monitoring techniques is presented in Chapter 4 and a number of case studies are provided in Chapter 5 which illustrate how remote sensing data can be applied. Digital image processing is a very important aspect of remote sensing and is conventionally dealt with in textbooks by showing the results of different procedures in plate format. However, it is essentially a practical hands-on process and a full understanding of the concepts can only be obtained by actually performing the processing. The CD that accompanies *Introductory Remote Sensing: Digital Image Processing and Applications* includes a modified version of DRAGON image processing software for PC. This will allow the reader to display imagery, histograms and scatterplots and to perform contrast stretching,

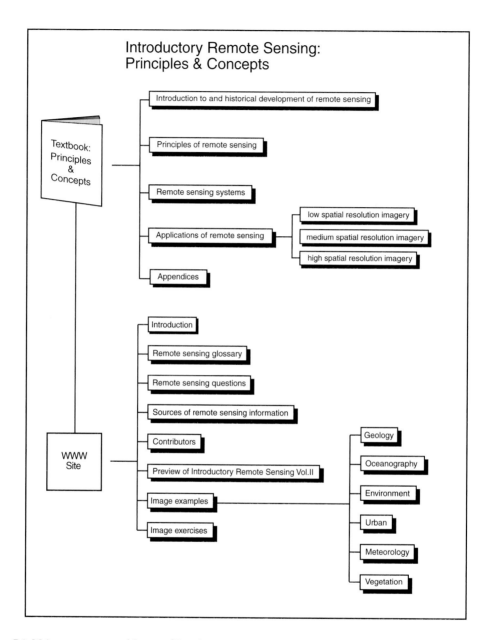

Figure P.1 Main components and layout of *Introductory Remote Sensing: Principles and Concepts.*

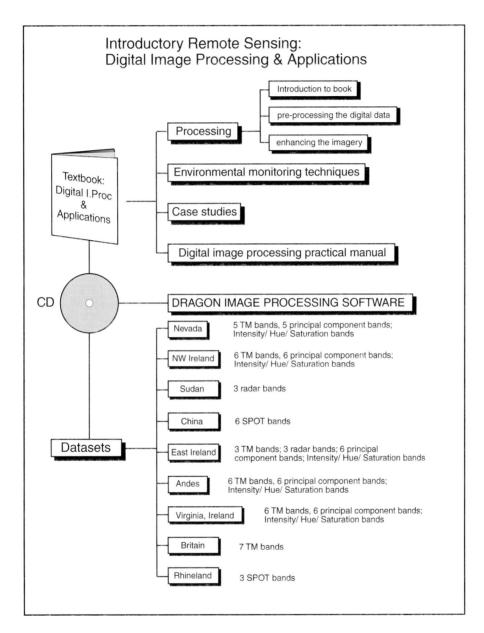

Figure P.2 Main components and layout of *Introductory Remote Sensing: Digital Image Processing and Applications.*

convolution filtering, ratioing and classification procedures. In all, 77 separate datasets are supplied on the CD for nine areas which encompass Europe, North America, South America, Africa and Asia. The datasets comprise Landsat TM, SPOT and radar imagery.

The format for both books is similar and is structured in such a way that:

- each chapter concludes with a short Further Reading section, though a much more comprehensive reading list is provided at the end of each book which allows the reader to deepen their knowledge;
- a short summary is provided for each chapter;
- a self-assessment test is included at the end of each chapter, in order that the readers may determine whether they have fully grasped the concepts discussed within the chapter. Answers for all the questions are provided in appendices for both books;
- a number of boxes are included within the main text, for example:

Example: Assuming Rayleigh scattering and using infrared radiation with a wavelength of 0.7 μm as a standard, calculate how much greater is the scattering for blue (0.4 μm); green (0.5 μm) and red (0.6 μm) light.

$$\text{Scattering for red light} = \frac{(0.7)^4}{(0.6)^4} = 1.85 \text{ times}$$

$$\text{Scattering for green light} = \frac{(0.7)^4}{(0.5)^4} = 3.84 \text{ times}$$

$$\text{Scattering for blue light} = \frac{(0.7)^4}{(0.4)^4} = 9.38 \text{ times}$$

These boxes relate to the material that is currently being explained and generally take the form of examples, including worked answers, which allow the readers to be confident that they have fully understood a particular point;

- shaded boxes are also included within the body of the text, for example:

Orthophotographs

An aerial photograph contains radial and terrain distortions. However, it is possible to correct these distortions, either by using a device called an orthophotoscope or by digital means. Such a corrected aerial photograph is termed an orthophotograph.

These shaded sections represent more complex components, and may be skipped if such in-depth knowledge is not required, or read in order to get a deeper insight.

Both *Introductory Remote Sensing* volumes may be used, either by an individual who wishes to gain a full understanding of the various concepts or in a classroom format. Volume I provides an excellent introduction to the subject and is particularly suited for fifth grade to first-year university level. The companion volume, which is more advanced, and requires a level of computer experience, may be used from sixth grade to third-year university level. The combination of the theory provided by the textbooks, exercises and examples on the dedicated WWW site and the practical component provided by the image-processing capabilities in the CD accompanying the companion volume, make this entire package particularly suited to teachers and lecturers who want to run a remote sensing course. With the increased application of remote sensing techniques in the fields of geography, geology, environmental monitoring, urban studies, oceanography and vegetation studies, the acquisition of remote sensing skills is extremely desirable and is expected to become increasingly important in the future.

ACKNOWLEDGEMENTS

I would like to thank Dr C. H. Power for her invaluable contributions to this text, especially Chapter 4, and Mr J. Keenan for his meticulous drawing of the figures within this book. Hilary Foxwell and Pat Brown also provided assistance. Grateful thanks are extended to Dr J. Keating for developing the Web product. A book such as this relies heavily on the goodwill of many individuals and organisations to provide relevant remote sensing imagery and data and the permission to use them. These organisations are listed in Appendix E but especial thanks go to Martin Critchley of ERA-Maptec and Hervé Lemeunier of SPOT IMAGE. A grant from the Publications Committee of the National University of Ireland, Maynooth, is gratefully appreciated.

1

INTRODUCTION TO AND HISTORICAL DEVELOPMENT OF REMOTE SENSING

1.1 INTRODUCTION TO REMOTE SENSING

Remote sensing can be defined as the *acquisition and recording of information about an object without being in direct contact with that object*. Although this definition may appear quite abstract and divorced from everyday living, most people have practised a form of remote sensing in their lives. A photograph obtained by a camera is a record which provides information about an object. A simple photograph such as that shown in Plate 1.1a can introduce us to some important remote sensing concepts. Different components within the photograph are different colours: the shirt is red, the trousers green and the folder blue.

This information about colour is carried by means of electromagnetic radiation and the colours represent specific ranges of the electromagnetic spectrum (see section 2.1). Spatial relationships between different components of a scene may also be determined from a photograph. On Plate 1.1a, the folder is held to the side of the figure at waist height and a building can be seen behind the person. However, in order to determine quantitative spatial information (for example, what is the area of the blue folder?), it is important that the scale of the photograph be known or an object of known length be included within the photograph. The metre rule held by the person allows the determination of distances and areas. Aspects of scale are considered in more detail in section 2.5.

Although the photograph shown in Plate 1.1a is a simple form of remote sensing, the images in this book have been acquired by airborne or spaceborne sensors and not by ground-based means. Remote sensing images are obtained at distances that fall within three broad ranges (Figure 1.1):

1 Sensors carried by aircraft generally obtain images at heights of 500 m to 20 km. In general most aerial surveys are carried out at heights of less than 5,000 m though some remote sensing experiments carried out in the United States of America have been typically performed at much higher altitudes (20 km) using converted U2 aircraft.
2 Sensors carried by spacecraft and satellites operate at distances of 250–1,000 km. Spacecraft (which are manned) generally operate at altitudes of around 250–300 km. The Russian government operates the MIR space station at altitudes of 300–400 km. Many remote sensing satellites (which are unmanned) operate approximately 1,000 km above the Earth. Examples of these types of space platforms are Landsat and SPOT (section 3.2). These systems provide data for a number of years and when they eventually fail, remain within their orbits whereas satellites in lower orbit may return to Earth in an uncontrolled manner and burn up as they pass through the atmosphere. Military surveillance satellites also operate within this 250–1,000 km zone.
3 Very high-altitude satellites operate 36,000 km above the Earth. These are geostationary satellites which have the unique capability of appearing to remain over the same part of the Earth at all times (section 3.3).

Plate 1.1b is a variation of the photograph shown in Plate 1.1a but it illustrates a very important concept as regards remote sensing: varying the distance at which images are acquired allows different types of information to be extracted. Detailed information about the person cannot be discerned on Plate 1.1b but his relationship to the surroundings can now be determined. The usefulness of this information is dependent upon the particular part of the scene that one is attempting to gain information on. There is therefore no single image which can be considered the 'best'; the optimum image is ultimately dependent upon the use that has to be made of it. Conversely, a close-up of Plate 1.1a would allow more information to be obtained on a small area, such as the blue folder, but at the expense of losing other information about the object. The ability to alter the distance from which the Earth is viewed and hence the resolution and the size of the area that is imaged is one of the most important attributes of remote sensing and allows it to be exploited in a wide variety of research fields. Small-scale images allow large areas of the globe to be analysed. Many current environmental concerns have a global dimension and are inextricably linked to climatic change.

The ability of satellites to provide a synoptic view of large parts of the Earth as an entity rather than in a piecemeal fashion is a major advantage. Such satellite images allow global problems to be viewed in a global context. Remote sensing should not be looked upon as a replacement for conventional field-based research but rather as a technique to complement it. 'Ground truth' is a commonly used (but misleading) term often applied to field investigations. Ground-collected verification data form an integral part of remote sensing. They provide an independent source of data and are important in the identification of features detected by remote sensing means. However, in many parts of the world ground-based research is minimal or researchers working far apart may not know of a common link between their researches. In addition, research is often confined to very narrow disciplines, often with little cross-referencing. Remotely sensed satellite images allow us to view the various components which comprise our environment (vegetation, cloud, rocks, cities) simultaneously. The interaction of many complex human-induced and natural processes produces the world that we live on today and global images allow us to analyse the interrelationship of these processes.

If remote sensing could only provide us with an overall view of the globe, while useful, its applications

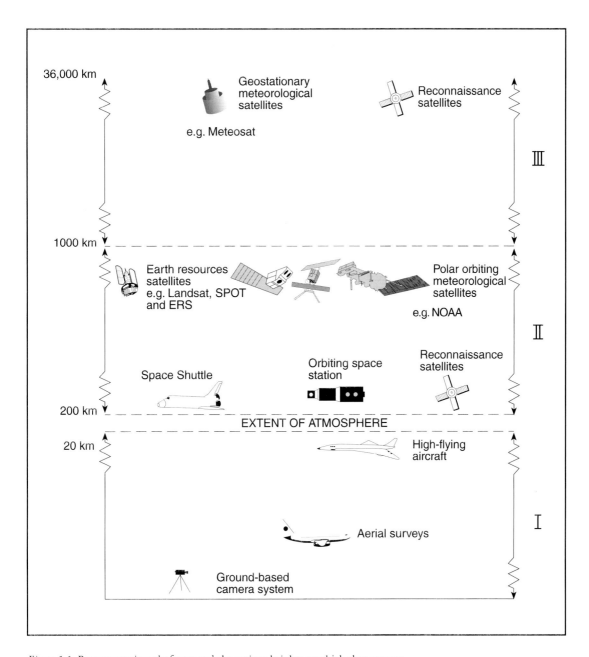

Figure 1.1 Remote sensing platforms and the various heights at which they operate.

would be limited. Cloud patterns can be determined on Plate 1.2a, but apart from the lakes in eastern Africa and the River Nile in northeast Africa, little information can be obtained on land areas. Plate 1.2a would not be particularly helpful if we were interested in studying the British Isles. However, we may view the same area from different heights and at different scales by using different remote sensing systems. The British Isles is much more recognisable in Plate 1.2b, which is an image obtained by another remote sensing system, though the definition is still poor. A more detailed view of northeast Ireland is illustrated in Plate 1.2c and allows much more information to be extracted for this smaller region. Rectangular field patterns emerge showing the cultivated areas; different colours can be related to vegetational variations. Red shows healthy vegetation, blue mostly bare soil and the very dark regular signatures are coniferous forest plantations divided up by linear tracks shown as white lines. (The colours on this image are 'false' as they do not accord with what we see with our eyes. See following paragraph.) An oblique aerial photograph of part of the coastline shown in Plate 1.2c is illustrated in Plate 1.2d, in which individual rock layers in the cliff can be discerned. The versatility to observe large areas or very small ones (i.e. varying the scale) makes remote sensing techniques applicable over a wide range of disciplines. An aerial photograph or a satellite image which shows a lot of detail (high spatial resolution) would be suitable for a forest manager who wishes to measure the area of forest whereas a low spatial resolution image, which provides information over a large area, is ideal for the climatologist.

Another important aspect of remote sensing demonstrated in Plate 1.1 has not yet been discussed. Three colours predominate in Plate 1.1a: blue, green and red. Every colour that can be observed by the human eye is formed by a combination of these colours. For example, yellow is formed by an equal combination of red and green with no blue input (see Plate 2.1a and Chapter 2 for a fuller discussion). Red, green and blue represent only a very small part of the electromagnetic spectrum. A comparison of Plates 1.2c and 1.2d shows that, whereas the colours observed on the latter image appear natural, vegetation has a bright red signature on the former image. This is because it is a false colour image which has been produced from data obtained in a part of the electromagnetic spectrum which cannot be detected by the human visual system. Two different surfaces may appear similar to the human eye but if they are viewed at a different wavelength, outside the visual

range, they can often be quite distinct. Computer technology allows us to 'see' wavelengths outside our normal visual range by using false colouration of these wavelengths. This ability to obtain information across a range of wavelengths, far greater than can be determined by the human eye, is another major advantage of remote sensing.

The production of an image such as that shown in Plate 1.2a is virtually an instantaneous record of the conditions that exist in a particular location at a specific time. If our environment was static this image would be a sufficient record of the events that occur at that location. However, the Earth is a dynamic system and natural forces are continually operating on it, causing it to evolve. Increasingly, human interference such as deforestation is accelerating these changes. Many remote sensing systems are designed to image the same location at periodic intervals; thus a record of the images obtained allows changes in the environment to be monitored and – more importantly – the rates at which these changes are occurring can be evaluated. An example of this important repeat facility is illustrated in Chapter 4, regarding the variation of ozone over Antarctica (see Plates 4.2a and 4.2b). The versatility shown by remote sensing systems regarding the area imaged and the wavelengths sensed, which has been discussed above, is mirrored in the different repeat cycles (i.e. how often an area is imaged). The Landsat system images most of the globe every 16 days and thus changes on a 16-day cycle can be determined. Such a repeat cycle is ideal for studying vegetational changes but would be entirely inappropriate for meteorological investigations because the atmospheric system alters much more rapidly. Consequently, meteorological satellites have much shorter repeat cycles, some as short as 30 minutes. When these properties of remote sensing systems are combined (the variation in area imaged and the detail obtained, the different wavelengths that can be sensed and the repetitive coverage that is available) and applied simultaneously in an investigation, then remote sensing techniques can be an extremely powerful approach for many applications which could not be attempted by conventional means. For example, using wavelengths at which vegetation is particularly sensitive, vegetational indices can be produced for entire continents. A gradual decline in a vegetation index over a number of years could be an indication that famine conditions are developing and following this warning, remedial action can be formulated and a possible catastrophe averted. Remote sensing is used extensively in Africa, where it is often the main or most reliable source of data. The Global and Early Warning System (GEWS) of the Food and Agriculture Organisation (FAO) in conjunction with the United States-funded Famine Early Warning System (FEWS) rely heavily on remotely sensed data for evaluating the possible development of famine conditions. Even if such a catastrophe occurs, remote sensing can still play an important role. During the drought and consequent famine in Eritrea in 1986, a team of researchers skilled in remote sensing techniques, based mainly at the Open University in the UK, analysed photographs of northeast Africa obtained by the Space Shuttle. The sites of potential sources of subsurface water were delineated from the photographs. The success rate for finding water by drilling at these specified locations was significantly greater than at other sites not identified from spaceborne sensors. Remote sensing has many applications, some of which are introduced in Chapter 4.

1.2 HISTORICAL DEVELOPMENT OF REMOTE SENSING

Remote sensing became possible with the invention of the camera in the nineteenth century. Astronomy was one of the first fields of science to exploit this technique and to this day much of astronomy is inextricably linked with remote sensing. A substantial amount of the progress in remote sensing has been as a result of its obvious military applications. Although information about the enemy was obtained from tethered balloons during the American Civil War, it was only during the First World War that free-flying aircraft were used in a remote sensing role. The static nature of that conflict meant that photographs taken from aircraft were an invaluable source of information

regarding the movement of troops and supplies, the reinforcement of fortifications and in assessing the effects of bombardments. Generally, breakthroughs in aspects of remote sensing such as sensor technology are a result of government-sponsored military research. Improvements are thus exploited initially by the military and – often much later – they filter down for civilian use. The development of remote sensing for specific applications has, in many cases, been a story of technological advance fired by political will rather than one of user and public demand creating the stimulus. The use of remote sensing for environmental assessment really became established after the Second World War. The two world wars had not only proved the value of aerial photography in land reconnaissance and mapping but had also driven technological advances in airborne camera design, film characteristics and photogrammetrical analysis. In peacetime these technological advances were devoted to civilian mapping and terrain assessment applications. However, up to the early 1960s, airborne missions were one-off, expensive surveys, providing data for relatively small areas at a single instant in time with no repeat coverage for comparative or change-detection purposes. Most countries have a national archive of aerial photographs, often held by their national mapping organisation. The majority are black and white, though in recent years colour has become more common. From about 1960, remote sensing underwent a major development when it was extended to space (Figure 1.2). Space rocket and satellite technology advanced greatly during the 1960s mainly because of the 'Space Race' between the USA and the former USSR. This new phase of remote sensing may be considered under four headings, though the sections are not mutually exclusive.

Military Reconnaissance Satellites

Prior to 1960, the United States and the then USSR exploited aerial photography as a means of gathering information about each other's military capabilities. However, in 1958 at the Surprise Attack Conference in Geneva, the western powers signalled the possibility of using satellites in order to identify armaments and their ancillary equipment which were greater than 25 m in size. The destruction of an American U2 'spy plane' by a missile while on a mission over the USSR in May 1960 provided added impetus to the development of space-monitoring capabilities and a few months after this incident the first of the reconnaissance satellites within the CORONA, ARGON and LANYARD programmes provided intelligence data. The satellites acquired photographic data at low altitudes compared with many satellites that are launched today (150–450 km). The duration of any one mission also tended to be very short (one to two weeks). This limitation was set by the amount of film that could be carried and the necessity to examine the photography as soon as possible. The film was ejected from the satellite in a canister and was recovered by specially equipped aircraft which intercepted the film canister as it descended to Earth. Initially cameras were flown singly but later satellites had two onboard cameras which allowed stereoscopic images to be obtained with a resolution of less than 2 m. (These reconnaissance satellites are often referred to as the Key Hole series, which is a designation based on the camera terminology.)

A 1970 photograph of part of Moscow is shown in Figure 1.3. The enlargement shows the Kremlin in some considerable detail. In 1995, many of these early photographs (those obtained between 1960 and

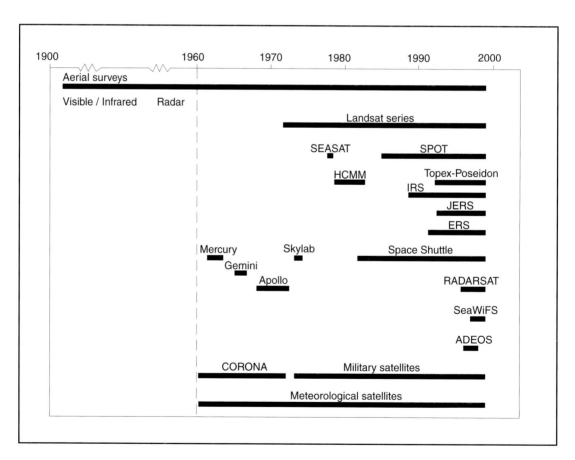

Figure 1.2 Operational dates for different remote sensing platforms. A marked change occurred about 1960 when spaceborne platforms became operational.

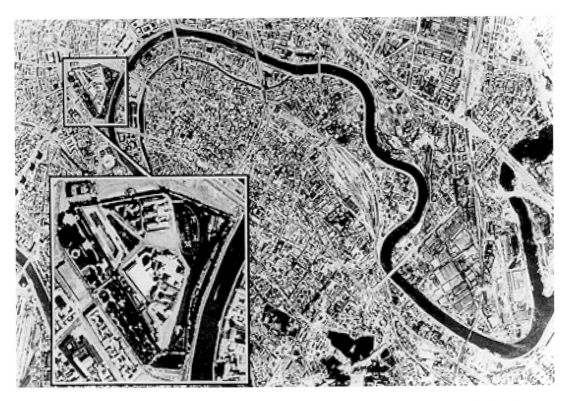

Figure 1.3 CORONA image of part of Moscow showing the Kremlin. Reproduced with permission, the American Society for Photogrammetry and Remote Sensing. McDonald, R. A., 'CORONA: success for space reconnaissance. A look into the Cold War and a revolution for intelligence', *Photogrammatic Engineering and Remote Sensing* volume 61, number 6, 1995, pages 689–720.

1972) were declassified and are now available to the public. Currently some American reconnaissance satellites orbit at approximately 700 km and may also incorporate an imaging radar system. Less is known about the Russian space programme but the Russians had comparable reconnaissance satellites, termed the KOSMOS series. The Russian authorities employed early reconnaissance satellites which were modified versions of the Vostok, Voskhod and Soyuz spacecraft. The craft were generally put into low Earth orbits of 100–250 km. The early satellites operated for only short periods, the duration of the missions being typically seven to eight days. Later generations could remain in orbit, providing data for hundreds of days. Some systems acquired their data in a digital rather than in a photographic format and these data were relayed to ground by telemetry.

Manned Space Flight

In 1961, the USA commenced its manned space flight programme which was to culminate in the first lunar landing in 1969. The first flights within the Mercury programme (1961–1963) lasted only a matter of hours. Photographs of the Earth were obtained by the astronauts through the capsule window. The acquisition of these data was not part of an experiment, however: these early photographs showed the potential for obtaining regional images of remote and inhospitable parts of the globe. The Gemini Project

(1965–1966) saw the first systematic approach to remote sensing from United States manned spacecraft when the Synoptic Terrain Photography experiment and the Synoptic Weather Photography experiment were conducted. Over 2,500 photographs were obtained using colour and infrared film. Although the Apollo missions (1968–1975) were primarily aimed at lunar landings, remote sensing of the Earth was performed while the spacecraft was orbiting the Earth. Between 1973 and 1974 three Skylab missions obtained over 44,000 images during the Earth Resources experiment. A range of remote sensing equipment was used during this experiment. Resolution for the images obtained from Skylab varied (60–145 m) depending on the system employed.

More sophisticated remote sensing systems were employed on Space Shuttle missions, which began in 1981. The Space Shuttle Earth Observation Program (SSEOP) is an ongoing programme which has obtained thousands of images to date. Images obtained from on board the Space Shuttle are shown in Chapter 3 and on the WWW site accompanying this book. Developments within the Russian space programme paralleled those of the United States and originally had a similar goal – to be the first country to place a person on the Moon. The first human in space was a Russian cosmonaut, Yuri Gagarin, who orbited the Earth on 12 April 1961. Although no photographs of the Earth were taken on this flight, information about what could be observed was recorded on a tape recorder. The Vostok programme (1961–1963) was analogous to the Mercury programme, involving a single astronaut, and the later Voskhod programme (1964–1965) was similar to the Gemini as it involved more than one astronaut in a space capsule. The Soyuz spacecraft essentially acted as supply craft and as a means for changing the crew for the Salyut space station, which the USSR began to develop after losing the undeclared race to get the first person on the Moon. The first Salyut station (19 April 1971) remained in orbit for only six months until 11 October 1971 and it was briefly inhabited for 20 days, during which time Earth photography experiments were performed. The Salyut space stations had military and civilian functions. Military

reconnaissance involved the use of a high-resolution camera with a spatial resolution of about a metre. Multispectral images with a resolution of 100 m were obtained in the visible and near infrared in nine separate wavebands. The Russians also carried out an Earth Resources Programme using a mapping camera (obtaining images for 450×450 km areas) and a multispectral camera with a 10–30 m resolution. Many of the Earth observation experiments were co-ordinated with aircraft using similar equipment which were obtaining images simultaneously in order that the photographs obtained in space could be compared with photographs obtained by airborne sensors.

Manned spacecraft have advantages and disadvantages as far as the acquisition of remote sensing data from space is concerned. The following are some of the advantages:

1. They are important testbeds for new systems. To send an unmanned satellite into orbit with a new and untried data-acquisition system may prove an expensive failure. However, the performance of a new system may be assessed in a manned flight and, in the light of the results obtained, may be altered to optimise its response before being committed to an unmanned satellite in orbit. For example, the success of the three radar-imaging experiments (SIR-A/B/C) performed on board the Space Shuttle in 1981, 1984 and 1994 was partly responsible for the decision to provide radar images from higher orbit satellites such as RADARSAT or ERS.
2. Trained personnel can rectify simple malfunctions on remote sensing systems which may be impossible to rectify on an unmanned one.
3. Human operators can take advantage of 'photographic opportunities' and can easily modify an existing programme in the light of changing circumstances. For example, a sudden volcanic eruption may be extensively analysed by being preferentially targeted.

The following are some of the disadvantages:

1. Manned space flights, whilst more flexible than unmanned missions, only occur at infrequent

intervals and for short durations. Thus the continual repetitive coverage produced by unmanned satellites provides much greater information.

2 Weight is a critical factor in any spacecraft launch and a large proportion of the weight of a manned spaceflight is taken up with life-support systems. In addition, fuel and engines must be retained for re-entry manoeuvres. An unmanned satellite may dispense with this equipment and either be lighter or include more remote sensing devices.

3 A malfunction aboard an unmanned satellite may impair that single operation but will not in general endanger life or the programme. The Challenger disaster in 1986, in which the astronauts were killed when the Shuttle exploded, effectively stopped flights for two years and the increased safety considerations have reduced the frequency of flights.

Meteorological Satellites

The weather forecast, and its associated images, is a daily routine for many of us and we now take for granted prior warnings about storms. Today, the movement of major hurricanes, the time and location of their expected landfall, the severity of the storm and deviations from its expected course can be broadcast on television and radio programmes almost continuously if required. However, prior to 1960, hurricanes and other tropical storms could often strike unsuspecting towns with little warning and with a subsequent loss of life and damage to property. Hurricanes form far out in the ocean and unless shipping provided information about the progress of these storms, little warning could be issued and minimal preventive action taken. The importance of having timely warning about major storms led to the launch by the United States of the first dedicated meteorological satellite on 1 April 1960. TIROS-1 (Television and Infrared Observation Satellite) was the first in a series of satellites which image large areas of the globe with a high repeat cycle.

Earth Resources Satellites

The applications of satellite Earth observations to environmental assessment and monitoring have been developing since the reception of the first meteorological satellite images in 1960. However, the lack of access to high-resolution data provided by the military reconnaissance satellites led to pressure from agencies like the United States Geological Survey for the formation of a civilian spaceborne remote sensing source of information. The first designated Earth Resources Satellite was launched in July 1972. The main sensors on board were particularly well suited for agricultural purposes. The satellite was originally termed ERTS-1 (Earth Resources Technology Satellite) but is now referred to as Landsat (see section 3.2). It was an experimental mission designed to acquire data from the Earth's surface on a systematic, repetitive, medium-resolution, multispectral basis. All data collected would be in accordance with the 'open skies principle', meaning that there would be non-discriminatory access to data collected anywhere in the world. All nations of the world were invited to take part in evaluating ERTS-1 data, and the results of the evaluation were overwhelmingly in favour of the continuation of the mission. This series of satellites has continued to the present day, though the onboard sensors have become more sophisticated. The first radar remote sensing satellite (SEASAT) whose data were available to the public was launched in 1978 but it provided data for only three months. Prior to the mid-1980s, the majority of satellites had been deployed by the USA and USSR. Information from the latter was not freely available (though it is now from the RESURS series) and a number of countries developed their own independent remote sensing satellites. France launched the first of the SPOT series in 1985 and in 1988 the first Indian Remote Sensing satellite (IRS) was put into orbit. Satellites launched by Japan include the Japanese Earth Resources Satellite (JERS) and the Marine Observation Satellite (MOS). China has been periodically putting satellites into orbit since 1975 though the data have not been freely available. Radar satellites have been launched in 1991 and 1995 by a European consortium

(European Radar Satellite – ERS) and by Canada in 1995 (RADARSAT). These satellite systems are considered in more detail in Chapter 3.

1.3 PHOTOGRAPHS AND IMAGES

The terms 'image' and 'photograph' are often used interchangeably in everyday conversation. However, in remote sensing these terms have more precise meanings. In the discussion of Plate 1.2, the term 'image' was applied to Plate 1.2a and 'photograph' to Plate 1.2d. A photograph is a representation of an object (or scene) that has been recorded on a film within the electromagnetic range of the spectrum encompassing the ultraviolet, visible and photographic infrared. A photograph is obtained by means of a camera system. An image is a representation of an object that is recorded by photographic means or by a scanner device. The scanner records digital numbers (DNs) which are related to a property of the object such as its reflectance. A scanner system can be designed to operate in particular ranges of the electromagnetic spectrum which cannot be sensed by a camera system as well as within the ultraviolet, visible and photographic infrared range to which photographs are restricted. When a collection of remotely sensed images and photographs is being considered, the general term 'imagery' is often applied.

1.4 CHAPTER SUMMARY

- Remote sensing can be defined as 'the acquisition and recording of information about an object without being in direct contact with that object'.
- remote sensing images can be obtained at different spatial scales, in different ranges of the electromagnetic spectrum and with different repeat cycles depending on the system that is employed.
- Before the 1960s remote sensing was confined to airborne platforms but since then many remote sensing systems have operated from space. Early

manned space flights showed the potential of images acquired from orbit and such images are still obtained today during Space Shuttle missions.

- Most spaceborne remote sensing data are obtained from unmanned satellites. Meteorological ones operate at heights of about 1,000 km or 36,000 km whereas Earth resources satellites generally operate at 1,000 km or less. Since 1972 a number of Earth resources satellites have been launched, most of which have resolutions of less than 100 m.

SELF-ASSESSMENT TEST

1 What are the major advantages of remote sensing for environmental monitoring?

2 What are the disadvantages of remote sensing?

3 Can scanner systems operate in the visible range of the electromagnetic spectrum?

4 Why were the 1960s so important for remote sensing?

5 At what heights do geostationary satellites operate?

6 What are the differences between images and photographs?

FURTHER READING

Calder, N. (1991) *Spaceship Earth*, London: Viking.

Drury, S. A. (1998) *Images of the Earth: a guide to remote sensing*, Oxford: Oxford Science Publications. (Chapter 1)

Furniss, T. (1986) *Manned Spaceflight*, London: Jane's Publishing Co.

Newkirk, D. (1990) *Almanac of Soviet Manned Space Flight*, Houston, Texas: Gulf Publishing Co. (Chapters 1 and 3)

Nicolson, I. (1982) *Sputnik to Space Shuttle*, London: Sidgwick and Jackson.

Porter, R. W. (1977) *The Versatile Satellite*, Oxford: Oxford University Press.

Verger, F. and Soubes, I. (1994) 'Evolution of remote sensing', *Sistema Terra* 3, 64–7.

2

PRINCIPLES OF REMOTE SENSING

2.1 ELECTROMAGNETIC RADIATION

In order to produce an image using data obtained by a remote sensing sensor, it is necessary to measure some parameter that can be related to the scene. A process must exist in order that this parameter can be conveyed to the sensor that may be thousands of kilometres away from the ground. For example, you can read this book because the information on the page is conveyed to your brain. This happens because visible light from the Sun (or an artificial lamp) hits the page and is then reflected or scattered into your eyes, which in this instance are analogous to remote sensing sensors. The white page reflects a substantial amount of the Sun's light and the dark words practically none. These signals, together with patterns and shapes, are decoded by the brain and interpreted as words. These signals are transmitted by means of electromagnetic waves and an understanding of their properties helps in the understanding of some remote sensing concepts.

Electromagnetic waves are transverse in character and consist of electric (E) and magnetic (H) fields that increase and decrease in phase with each other (Figure 2.1). The electric and magnetic fields are always at right angles to each other and to the direction in which the wave propagates. Transverse waves have a number of characteristics, of which the most important are wavelength, frequency and amplitude. The wavelength (λ) is the distance between two adjacent peaks (Figure 2.1). The wavelengths sensed by many remote sensing systems are extremely small and are measured in terms of micrometres (μm: 10^{-6} m) or nanometres (nm: 10^{-9} m). Frequency (f) is defined as the number of peaks that pass any given point in one second and is measured in Hertz (Hz). Amplitude is the maximum value of the electric (or magnetic) fields and is a measure of the amount of energy that is transported by the wave. A comparison of the waves shown in Figure 2.1 shows that an inverse relationship exists between wavelength and frequency. The longer wavelength wave (λ_1) has a lower frequency than the shorter wavelength wave (λ_2). All matter at a temperature greater than 0 K ($-273\,^\circ$C) produces electromagnetic radiation and all electromagnetic radiation travels at the same speed in a vacuum. This speed, termed the speed of light (c), is approximately 3×10^8 m/s. The relation between λ and f is given by:

$$c = f\lambda \qquad \text{equation 2.1}$$

Example: Calculate the frequency of electromagnetic radiation with a wavelength of 700 nm.

$$c = f\lambda$$

therefore $f = \dfrac{c}{\lambda}$

$$f = \frac{3 \times 10^8}{700 \times 10^{-9}}$$

$$= 4.28 \times 10^{14}\,\text{Hz}$$

Although visible light is the most obvious manifestation of electromagnetic radiation, other forms also exist. If an electric iron is switched on, its visible appearance does not change, but if your hand is held near the base plate it becomes warm. The transfer of energy to your hand shows that electromagnetic radiation at a wavelength not within the visible range is being produced. In this instance the electromagnetic radiation is thermal radiation, which is at a longer wavelength than visible light. Electromagnetic radiation can be produced at a range of wavelengths that encompass many orders of magnitude (Figure 2.2). By convention this continuous spectrum is divided into discrete regions though in reality the boundaries indicated are often gradational and may overlap because some of the subdivisions of the electromagnetic spectrum are based on how the radiation is generated. Gamma rays, X-rays and ultraviolet light operate at shorter wavelengths than visible light whereas infrared, microwave and radio waves operate at longer wavelengths. (However, it is important to remember that they all travel at the same speed in a vacuum.) Visible light occupies only a minute fraction of the available electromagnetic spectrum. However, the full range of the spectrum is not available for remote sensing purposes. Only the higher range of

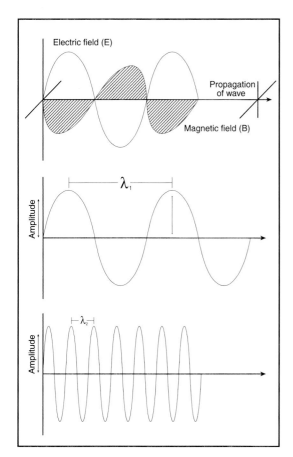

Figure 2.1 Propagation of electromagnetic energy by means of a transverse wave. The waves may have different amplitudes, frequencies and wavelengths.

Example: Using Figure 2.2, what percentage does the visible light range cover if the available range extends to 30 cm?

Available range = 30 cm = 30×10^{4} μm
Visible range = 0.4 μm to 0.7 μm = 0.3 μm

Therefore percentage of range covered by visible light:

$$\frac{0.3}{30 \times 10^{4}} \times 100$$

$$= 10^{-4} \text{ per cent}$$

the ultraviolet, all the visible and parts of the infrared and microwave ranges are employed for remote sensing purposes (see section 2.2). Even within this more restricted range, visible light occupies a minor part.

Visible light only encompasses one ten-thousandth of 1 per cent of the possible range. The visible range is further subdivided into the blue, green and red ranges, where each colour represents a specified waveband (Figure 2.2). Mention the word 'infrared' to an average person and he/she will immediately think of heat. However, it is important to realise that a significant amount of remote sensing performed within the infrared waveband is not related to heat. This is the photographic infrared range that is at a slightly longer wavelength than red. Although this part of the infrared is invisible to the human eye, it can be detected by photographic film. Thermal infrared remote sensing is carried out at longer wavelengths but not by photographic means; see section 3.1.

Polarisation of Light

Figure 2.1 shows a single electromagnetic wave in which the electric field vector (E) is increasing and decreasing in a vertical plane. However, natural electromagnetic radiation coming from the Sun can be thought of as consisting of transverse waves which increase and decrease in amplitude with random orientations. Such radiation is said to be unpolarised. It is possible to process the unpolarised light in such a way that all the random waves are blocked off except for a wave that is vibrating with a particular orientation. Electromagnetic radiation, which is designed to increase and decrease in one plane only, is termed 'plane-polarised' (or more simply 'polarised'). Polarisation effects are generally not taken account of for remote sensing operating at visible and near-infrared wavelengths. However, at longer wavelengths, for example in the microwave range where remote sensing systems such as radar operate, the generated electromagnetic radiation is polarised within a certain plane and the returned radiation is also measured in a certain orientation.

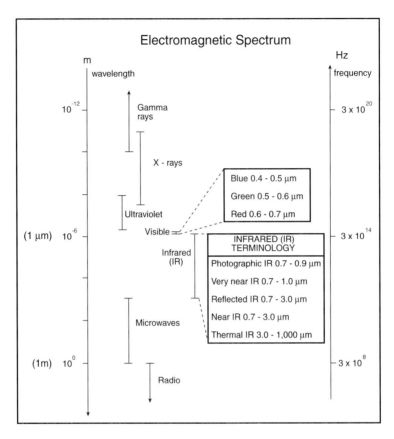

Figure 2.2 Nomenclature of the electromagnetic spectrum. Note the overlap between some of the bands such as between X-rays and gamma rays and also with regard to the infrared terminology.

Characteristics of Solar Radiation

Passive remote sensing uses the Sun as its source of electromagnetic radiation. The characteristics of the Sun's radiation are therefore of importance as they will have some bearing on the techniques that are employed and the sensors that have been developed for remote sensing purposes. The Sun has a surface temperature of 5,750–6,000 K and radiates energy across a range of wavelengths at an average distance from the Earth of 150 million kilometres. This transference of energy to the Earth occurs almost exclusively in a near vacuum.

For this distance, apart from a massive decrease in intensity across all wavelengths (only 5×10^{-9} per cent of the Sun's output reaches the Earth), there is little change in the proportions of each waveband. However, to reach the Earth's surface the radiation must pass through a 100 km thick atmosphere and within this narrow zone the electromagnetic radiation is selectively absorbed and scattered (see section 2.2). Although the Sun produces electromagnetic radiation across a range of wavelengths, the amount of energy it produces is not evenly distributed (Figure 2.3a). Ninety-nine per cent of the energy is within the 0.28–4.96 μm waveband. Approximately 43 per cent is radiated within the visible waveband from 0.4 to 0.7 μm. At wavelengths shorter than the visible range (gamma rays, X-rays and ultraviolet) about 9 per cent is produced, mainly within the ultraviolet. Forty-eight per cent of the energy is transmitted at wavelengths greater than 0.7 μm, mainly within the infrared range, with little energy in the microwave part of the electromagnetic spectrum.

A blackbody can be defined as a body that emits the maximum intensity of radiance across all wavelengths that it is theoretically possible to radiate for the particular temperature of the body. It absorbs and re-emits all the energy incident on it. Although this is a theoretical concept, the Sun and the Earth are often modelled on the assumption that they are blackbodies. The energy emitted from a blackbody is given by the Stefan– Boltzmann Law:

$$M = \sigma T^4 \qquad \text{equation 2.2}$$

where M is the energy emitted; T is the temperature in degrees Kelvin and σ is the Stefan–Boltzmann constant which is equal to 5.6697 $\times 10^{-8}$ W m^{-2} K^{-4}.

Example: Calculate the temperature of a blackbody if the measured energy output is 3,544 W m^{-2}.

From equation 2.2: T^4

$$= 3544/5.6697 \times 10^{-8}$$

$$= 625 \times 10^8$$

Therefore T equals the fourth root of 625×10^8

thus T = 500 K

Although, intuitively, we accept that the Sun emits energy, equation 2.2 shows us that all matter at a temperature greater than 0 K radiates energy. Thus a block of ice, which we perceive as being cold, also emits electromagnetic radiation. The power of 4 in equation 2.2 shows that the emitted energy increases substantially even for relatively small temperature rises. The Earth also radiates energy but at an average temperature of 300 K. Thus the Sun emits over 160,000 times more energy per unit area than the Earth ($6,000^4$/300^4). The wavelength at which a blackbody radiates its maximum energy is inversely proportional to temperature (Figure 2.3b) and is given by Wien's displacement law:

$$\lambda_m = \frac{A}{T} \qquad \text{equation 2.3}$$

where λ_m is the wavelength at which the maximum emittance is produced, A is a constant = 2,898 μm K and T is the temperature in degrees Kelvin. Note that because the units of A are given in micrometres the wavelength will also be in micrometres. The inverse relationship between λ_m and T can be demonstrated by heating a bar of metal. At low temperatures the bar will emit heat but the appearance of the metal will not change because the energy is being emitted within the thermal range of the electromagnetic spectrum. As the temperature increases, the metal will initially glow a dull red and then yellow, which signifies a shift in the maximum energy output to lower wavelengths. Using the surface temperatures for the Sun and the Earth and substituting them into equation 2.3 shows that the Sun's maximum output is at a wavelength of about 0.5 μm and the Earth radiates its maximum energy at a much longer wavelength (9.66 μm) which cannot be detected by the human eye because it is within the thermal infrared range of the electromagnetic spectrum.

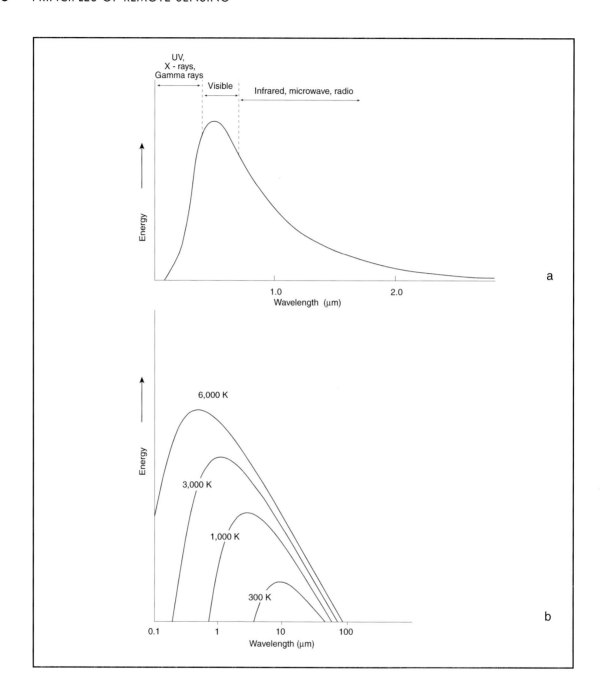

Figure 2.3 (a) Variation of intensity of radiation with wavelength for the Sun. The maximum energy is emitted within the visible range. (b) The wavelength at which the maximum energy emitted is inversely proportional to temperature.

Passive and Active Remote Sensing Systems

There are two basic types of remote sensing system: passive and active. A passive system uses the Sun as the source of electromagnetic radiation. Radiation from the Sun interacts with the surface (for example by reflection) and the detectors aboard the remote sensing platform measure the amount of energy that is reflected (Figure 2.4). An active remote sensing system carries onboard its own electromagnetic radiation source. This electromagnetic radiation is directed at the surface and the energy that is scattered back from the surface is recorded (Figure 2.4). A camera is normally a passive means of recording data. If the photograph is obtained in daylight, the electromagnetic radiation within the visible waveband is reflected from the object and recorded by the film. However, if the photograph is obtained at night, the camera becomes an active remote sensing system. An electronic flash is initially fired which directs an electromagnetic pulse at the target in order to illuminate it and the reflected radiation is received and recorded back at the camera. The commonest active remote sensing system is radar that produces electromagnetic

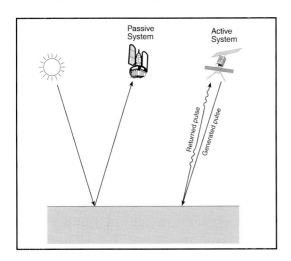

Figure 2.4 Active and passive remote sensing systems. A passive system uses the Sun as the source of electromagnetic radiation whereas an active system, such as radar, produces its own electromagnetic radiation.

radiation in the microwave band of the spectrum. Other active systems that have been used in remote sensing are lasers and sonar devices. However, in the latter case, the energy waves are not electromagnetic but longitudinal in nature and do not travel at the speed of light.

2.2 CHARACTERISTICS OF THE ATMOSPHERE

Composition of the Atmosphere

The composition of the atmosphere is of importance in remote sensing because electromagnetic radiation must pass through it in order to reach the Earth's surface. The atmosphere is held close to the Earth by the force of gravity. The upper boundary marking the outer limit of the atmosphere is not sharply defined but only 3×10^{-5} per cent of the gaseous atmospheric components is found above 100 km.

Nitrogen and oxygen combined make up 99 per cent of the atmospheric gases in a ratio of 4:1 (Table 2.1). Argon, an inert gas, forms just less than 1 per cent. Most of the ozone is concentrated within the stratosphere between 19 km and 23 km. The atmosphere also contains other molecular species such as water vapour (H_2O) and methane (CH_4). The size of the molecules varies but they typically have dimensions of 10^{-4} μm. The atmosphere also contains particles with a range of sizes and sources which are of great importance in remote sensing. Dust particles have radii that vary from 0.01 to 10 μm and form about 25 per cent of the total number of particles in the atmosphere. Sources include windblown dust

Table 2.1 Gaseous composition of the atmosphere

Component	Percentage
N_2	78.08
O_2	20.94
Ar	0.93
CO_2	0.0314
O_3	0.00000004

from desert areas and volcanic ash. Pollen from vegetation also represents approximately 25 per cent of the particle atmospheric component and ranges in size from 10 to 100 μm. Smoke particles are typically 0.05–1 μm in size and may be caused by natural forest fires or produced anthropogenically. The components of the atmosphere play a two-fold role that is important for remote sensing: absorption and scattering.

Absorption in the Atmosphere

Although electromagnetic radiation of all wavelengths emitted by the Sun reaches the top of the atmosphere, only radiation within specific wavebands can pass through the atmosphere to reach the surface of the Earth. This is because the gaseous components of the atmosphere act as selective absorbers. Different molecules absorb different wavelengths. The arrange-

ment of the molecules and the energy levels of the specific atoms in the gases largely determine what wavelengths are absorbed. Nitrogen, the commonest gaseous component of the atmosphere, has no prominent absorption features apart from an absorption band at wavelengths less than 0.1 μm. Oxygen absorbs in the ultraviolet and also has an absorption band centred on 6.3 μm. Carbon dioxide prevents a number of wavelengths reaching the surface. A broad absorption band exists between 14 and 17 μm and narrower ones occur at 2.7 μm and 4.5 μm (Figure 2.5a). Water vapour is an extremely important absorber of electromagnetic radiation within the infrared part of the spectrum. Absorption bands also exist at 1.4 μm, 2.7 μm and 6.3 μm. Ozone (O_3) has an important absorption band within the ultraviolet, hence the concern about the depletion of this gas in the atmosphere. A reduction in atmospheric ozone will allow more ultraviolet radiation to reach the

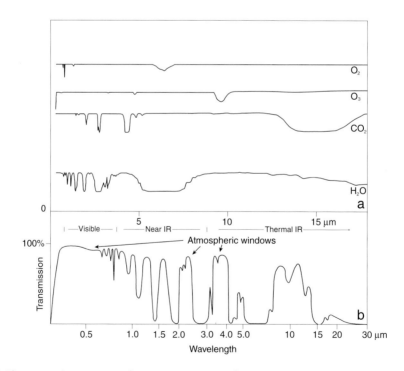

Figure 2.5 (a) Absorption characteristics of gaseous components of the atmosphere. (b) Transmission through the atmosphere as a function of wavelength and the location of atmospheric windows. Modified from Sabins (1997).

Earth's surface, with a possible increase in skin cancer. The gases in the atmosphere do not act in isolation. The combined effects of the absorption characteristics of the atmospheric gases mean that:

1 electromagnetic radiation at particular wavelengths is totally absorbed and does not reach the Earth's surface;
2 electromagnetic radiation at particular wavelengths is partially absorbed and only a proportion that reaches the Earth's outer atmosphere passes through the atmosphere to reach the ground;
3 electromagnetic radiation at particular wavelengths is unaffected by atmospheric absorption

and virtually all that reaches the Earth's outer atmosphere reaches the surface.

The wavelengths at which electromagnetic radiation are partially or wholly transmitted through the atmosphere are known as atmospheric windows (Table 2.2 and Figure 2.5b).

The major atmospheric windows occur:

1 within the visible and photographic infrared range where there is approximately 95 per cent transmission of the radiation across a broad atmospheric window. The window extends into the higher wavelength sections of the ultraviolet and thus encompasses a total waveband of 0.3–1.0 μm;

Carbon Dioxide and the Greenhouse Effect

The Sun has been continually bombarding the Earth with electromagnetic radiation for 4 billion years and yet its average temperature is only 27 °C. This indicates that the Earth is also losing heat and that over geological time an equilibrium has been achieved. It is possible to calculate the temperature that the Earth should be in order to achieve this balance and it is about 40 °C lower than the average temperature. The much higher temperature that the Earth currently experiences is caused by absorption within the atmosphere. The Earth is heated by short wavelength radiation, but it re-radiates electromagnetic radiation at much longer wavelengths (the maximum emittance is at about 9.7 μm). The atmosphere absorbs this long-wavelength radiance and re-emits it, heating the Earth up in the process. This greenhouse effect, where longer wavelength energy is absorbed by the atmosphere, is vital for the existence of life on Earth. However, concern has been expressed over the increase in the Earth's temperature resulting from the introduction, over the past 50 years, of gases into the atmosphere at an ever-accelerating rate which may disrupt the nat-

ural balance. Concern has focused on carbon dioxide emissions, which are released by the burning of fossil fuels. There are two main reasons why concern about the concentration of carbon dioxide in the atmosphere is growing.

1 Figure 2.5a shows that carbon dioxide has a strong absorption band at 14–17 μm which is close to the wavelength at which the maximum energy is emitted from the Earth (9.7 μm). Thus carbon dioxide is an efficient absorber of radiation at the wavelength at which most of the Earth's radiation is emitted.
2 If carbon dioxide formed a large proportion of the atmosphere, any carbon dioxide released by humans would be a tiny fraction of the natural background concentration and would have very little effect on temperatures. However, carbon dioxide represents only 0.0314 per cent of the atmosphere. Consequently, humans have the capacity to alter this small proportion significantly. Estimates vary but it is believed that over the last 200 years there has been a 30 per cent increase in atmospheric carbon dioxide.

Table 2.2 Wavelength of atmospheric windows

Atmospheric window	Waveband (μm)
Upper ultraviolet – photographic infrared	0.3–1.0 (approx.)
Reflected infrared	1.3, 1.6, 2.2
Thermal infrared	3.0–5.0
Thermal infrared	8.0–14.0
Microwave	> 5,000

2   within the reflected infrared part of the electro-magnetic spectrum centred on specific narrow wavebands, 1.3 μm, 1.6 μm and 2.2 μm;

3   within two broad bands in the thermal infrared at 3–5 μm and 8–14 μm;

4   wavelengths greater than about 0.5 cm because the atmosphere is transparent to electromagnetic radiation at microwave wavelengths. To obtain information in this part of the spectrum, either very small signals must be measured or an artificial source of microwaves is required. The latter approach involves the use of active radar systems.

The sensors on remote sensing systems must be designed in such a way as to obtain their data within these well-defined atmospheric windows. However, we have not yet considered the differences in absorption by atmospheric gases between an airborne sensor and a spaceborne sensor. Consider the situation shown in Figure 2.6. Electromagnetic radiation is partially absorbed by the atmosphere but sufficient reaches the surface from which it is reflected and a sensor aboard the aircraft measures this signal. In order for the spaceborne sensor to record the same electromagnetic radiation, the radiation has to travel though the atmosphere to reach the Earth's surface and then travel through the entire atmosphere again to be recorded by the sensors on board the satellite – a two-way transit. As a result virtually no signal may be obtained at the spaceborne sensor. (Note: if the atmosphere were 100 per cent transparent at this wavelength then the satellite would be able to record the radiation.)

Scattering in the Atmosphere

Scattering of electromagnetic radiation by aerosols in the atmosphere is a phenomenon that we experience in our everyday lives. When you walk into the shadow of a building out of direct sunlight, although you are now in the shade, you are able to see because of scattering in the atmosphere. This is because incoming radiation is scattered by the aerosols and indirectly provides illumination (Figure 2.7a). The scattering mechanisms can be selective or non-selective. In selective scattering, the relative size of the particles in the atmosphere and the wavelength of the electromagnetic radiation is important whereas in non-selective scattering the dimensions of the particles and the wavelengths are not relevant.

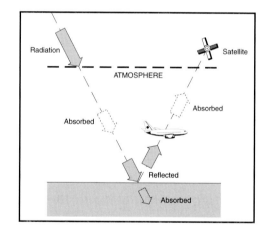

Figure 2.6 Radiation measured by a satellite system has a two-way transit through the atmosphere compared with an aerial system. Thus complete absorption of a particular wavelength may result in the satellite system recording no data at that wavelength.

Selective Scattering

Rayleigh scattering (also termed molecular scattering) occurs when the dimensions of the scatterers are small (less than one-tenth the size) compared with the wavelengths of the electromagnetic radiation. Molecules of oxygen and nitrogen (commonest constituents of the atmosphere) fulfil this role for visible radiation. The amount of scattering is inversely proportional to the fourth power of the wavelength.

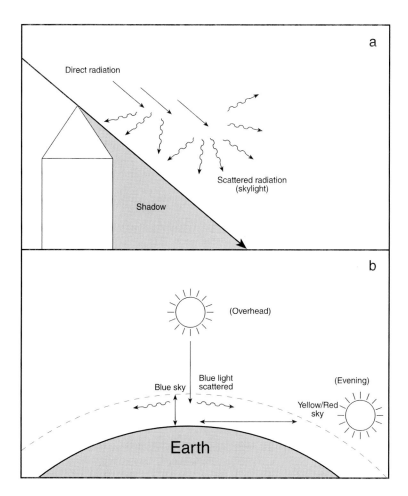

Figure 2.7 (a) Scattering in the atmosphere indirectly illuminates areas in shadow. (b) Scattering of short wavelength blue light results in a blue sky. At dusk and dawn when the atmospheric path for radiation is longer, longer wavelengths may be scattered, yielding a red/orange sky.

Within the visible range of the electromagnetic spectrum, blue light is scattered to a much greater degree than green or red. This is the principal reason why the sky appears blue – this waveband is scattered in all directions including towards the surface. As one moves vertically up through the atmosphere, it becomes darker because there are fewer scattering aerosols present at higher altitudes. At sunset we often see spectacular reds and yellows in the sky. This is because the radiation has had to travel through a greater thickness of atmosphere, causing a greater degree of scattering, and thus most of the shorter wavelength radiation has been scattered away and the longer wavelengths' colours reach our eyes (Figure 2.7b).

When the dimensions of the aerosols in the atmosphere are approximately the same as the wavelength of the electromagnetic radiation, then Mie or non-molecular scattering occurs. Aerosols such as dust, smoke and the smallest pollen grains contribute to Mie scattering but gas molecules are too small to produce Mie scattering at the wavelengths used for remote sensing. Mie scattering is also wavelength dependent and varies approximately as the inverse of the wavelength.

Example: Assuming Rayleigh scattering and using infrared radiation with a wavelength of 0.7 μm as a standard, calculate how much greater is the scattering for blue (0.4 μm); green (0.5 μm) and red (0.6 μm) light.

$$\text{Scattering for red light} = \frac{(0.7)^4}{(0.6)^4} = 1.85 \text{ times}$$

$$\text{Scattering for green light} = \frac{(0.7)^4}{(0.5)^4} = 3.84 \text{ times}$$

$$\text{Scattering for blue light} = \frac{(0.7)^4}{(0.4)^4} = 9.38 \text{ times}$$

Non-Selective Scattering

Scattering at all wavelengths occurs equally with aerosols whose dimensions are greater than approximately ten times the wavelength of the radiation. For visible wavelengths (0.4–0.7 μm) the main sources of non-selective scattering are pollen grains, cloud droplets, ice and snow crystals and raindrops. Many of the equations derived for scattering assume that the relevant particles are spherical in shape. However, in reality the particles are often quite irregular. In addition, they often carry electric charges and may coalesce to form larger particles with more complex shapes. Thus the equations of scattering are useful for indicating general processes though the actual scattering mechanisms that occur are very complex. Upon traversing the Earth's atmosphere, the electromagnetic radiation, which has by now been selectively scattered and absorbed, reaches the Earth's surface.

2.3 INTERACTION OF ELECTROMAGNETIC RADIATION WITH A SURFACE

When electromagnetic radiation strikes a surface, it may be reflected, scattered, absorbed or transmitted (Figure 2.8). These processes are not mutually exclusive: a beam of light may be partially reflected and partially absorbed. Which processes actually occur depends on the wavelength of the radiation, the angle at which the radiation intersects the surface and the roughness of the surface. Reflected radiation is returned from a surface at the same angle as it approached; the angle of incidence thus equals the angle of reflectance. Scattered radiation, however, leaves the surface in all directions. The concept of scattering is often subsumed within reflection and it is termed 'diffuse reflection'. Whether or not incident energy is reflected or scattered is partly a function of the roughness variations of the surface compared to the wavelength of the incident radiation. If the ratio of roughness to wavelength is low (less than one), the radiation is reflected whereas, if the ratio is greater than one, the radiation is scattered. A surface which reflects all the incident energy is known as a specular reflector whereas one which scatters all the energy equally is a Lambertian reflector. Real surfaces are neither fully specular nor Lambertian; however, in general, the more Lambertian a surface is, the better it is for remote sensing purposes. If images are acquired by an aircraft flying over a uniform Lambertian surface, the reflectance obtained for the surface at any particular wavelength will be similar irrespective of the location of the aeroplane because the radiation is being scattered equally in all directions. However, if the surface is wholly specular, a bright signature would be obtained for one position of the aircraft and dark signatures for the same surface at other positions. This widely varying signature for the same feature would significantly reduce the effectiveness of remote sensing for monitoring purposes and make statistical analysis and classification procedures much less reliable. Most natural surfaces that are observed using remote sensing systems are approximately Lambertian at visible and infrared wavelengths. An exception to this rule is water. Figure 2.9 shows a number of small lakes in which water has a different signature. The tone for the water is generally very dark, but the small lake near the eastern edge of the photograph is associated with a pale signature (high reflectance due to 'sun glint'). A further complication is introduced by the action of wind on the water. This may generate surface waves that

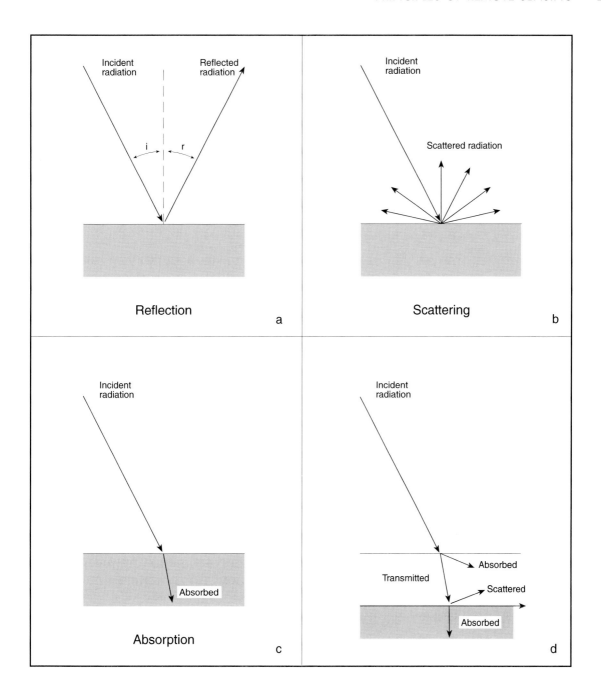

Figure 2.8 Interaction of electromagnetic radiation with a surface. The radiation can be (a) reflected, (b) scattered, (c) absorbed or (d) transmitted.

will have a rougher texture than calm water and also alter the angle of the incident radiation. The long narrow lake shown in Figure 2.9 has a different signature at its southern end because of the production of wind-generated waves.

The albedo of a surface is given by the ratio of the electromagnetic radiation reflected from a surface to the total electromagnetic radiation incident on the surface (Table 2.3). The Earth, including the atmosphere, has a global albedo of around 34 per cent, about 75 per cent of it caused by reflection from clouds. The large variation for water is due to the

Table 2.3 Albedo of various surfaces

Surface type	Albedo (%)
Grass	25
Concrete	20
Water	5–70
Fresh snow	80
Forest	5–10
Thick cloud	75
Dark soil	5–10

angle at which the incident energy hits the surface.

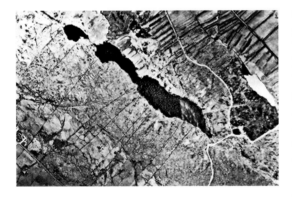

Figure 2.9 Reflectance characteristics for water. Water generally has a low reflectance and thus a dark signature. However, when it acts as a specular reflector, as in the eastern part of the image, it may have a high reflectance. (North is always at the top of the images.) Based on the Ordnance Survey of Ireland by permission of the government (permit number 6259).

Water tends to have a lower albedo at lower incidence angles.

A proportion of the electromagnetic radiation incident on the Earth's surface is absorbed. This energy is then available to be emitted at longer wavelengths and can be measured by sensors that are tuned to the thermal infrared. A component may also be transmitted through the surface if it is a medium such as water. This transmitted energy may eventually reach the bottom of the water column if it is not too deep and be scattered or absorbed (Figure 2.8d).

Various estimates have been put forward regarding the percentage of electromagnetic radiation that traverses the atmosphere to reach the surface of the Earth to be scattered, absorbed or reflected and finally measured by remote sensing sensors (Figure 2.10). Approximately half of the incident energy reaches the Earth's surface and is absorbed, whereas only 4 per cent is reflected and scattered from the surface. Approximately 6 per cent is scattered into space or towards the surface by the atmosphere and a further 20 per cent is reflected from clouds, which also absorb 5 per cent.

Spectral Signatures of Landscape Features

Our experiences in everyday life tell us that different features are different colours: grass is green, sky is blue and so on. Grass is green to our eyes because it has a higher reflectance in green than in blue or red. The reflectance of a body is wavelength dependent and it may change markedly over a range of few micrometres (see Figure 2.11, for example). Remote sensing systems often operate at wavelengths which cannot be detected visually, and in order to understand the signatures obtained by their sensors we require a knowledge of the reflectance and absorption properties of different features that make up a landscape. A graphical representation of the reflectance variations as a function of wavelength is known as a spectral reflectivity curve. Discussion in this section refers mainly to the visible to near infrared range. Discussion of the thermal and the microwave range of the electromagnetic spectrum is deferred until

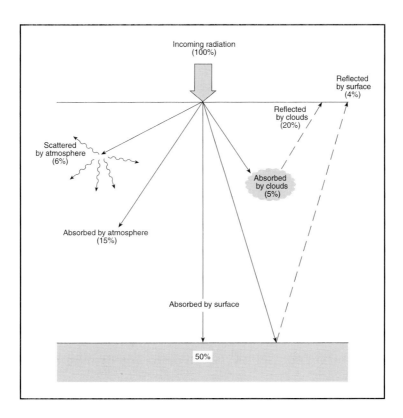

Figure 2.10 Interaction of electromagnetic radiation with the atmosphere, which selectively absorbs and scatters the radiation such that only a small component is reflected from the Earth's surface and measured by remote sensing sensors.

Chapter 3 when thermal and radar images are considered. The characteristics of a scene, which may be imaged by remote sensing sensors and which are discussed here, are the spectral signatures of:

- vegetation and soil;
- water (liquid and solid phase);
- rocks;
- cultural features.

Vegetation

The reflectance and absorption characteristics of vegetation are complex because vegetation has seasonal cycles; the signature obtained will thus vary greatly depending on the particular phase of the cycle. Electromagnetic radiation can also be transmitted through vegetation, where it may interact with the constituent parts. Figure 2.11 shows a generalised reflectivity curve for green vegetation. Within the visible range, green vegetation has an absorption band in the blue and red parts of the spectrum because of the presence of chlorophyll. Even within the green part of the spectrum (where the highest reflectances in the visible range occur), only 10–15 per cent of the incident light is reflected. In general vegetation is a good absorber of electromagnetic radiation at visible wavelengths. A marked change in the absorption/reflectance properties occurs at the red/infrared boundary around $0.7\,\mu m$ where absorption by the constituents of vegetation is greatly reduced and reflectance increases greatly. Reflectance is approximately constant between $0.7\,\mu m$ and $1.3\,\mu m$ but decreases at longer wavelengths. Troughs in the spectral reflectivity curve centred on $1.4\,\mu m$

Leaf Structure and Reflectance from Layered Canopies

The spectral response of vegetation primarily depends upon the structure of plant leaves. The leaf consists of layers composed of different types of cells (Figure 2.12). The upper or outer layer is formed of transparent protective epidermis cells that can be penetrated by all wavelengths of electromagnetic radiation. The next layer (palisade cells) contains sacs of green pigment (chlorophyll) called chloroplasts. This layer of cells is responsible for the green appearance of healthy living vegetation. All the colours of visible electromagnetic radiation except green are absorbed by the chloroplasts; the green is reflected back to our eyes or towards the satellite sensor, and the leaf therefore appears green. Between 10 and 30 per cent of the total amount of visible light arriving at the surface of the leaf is reflected back as green. Other wavelengths of electromagnetic radiation are absorbed by this layer too including the infrared. However, the larger irregular packing cells, which make up the body of the leaf (mesophyll cells), reflect about 60 per cent of the near-infrared radiation reaching this leaf layer. Healthy vegetation therefore has a higher or brighter response in the near infrared than in the green part of the spectrum. Another thin layer of lower epidermis cells can be found beneath the mesophyll cells on the underside of the leaf. Healthy live vegetation can be discriminated from dead vegetation by the use of both the visible and infrared parts of the spectrum. As a leaf dies, cells within it die, causing the palisade cells to lose their green pigment. Red and blue light therefore are no longer absorbed by these cells and are reflected back along with the green, and thus dead and dying vegetation appears yellow or brown. Near-infrared wavelengths are no longer reflected but are absorbed by the dead mesophyll cells, so that dead vegetation appears dark or black in the near infrared. Such changes are used to monitor the health of vegetation.

Although some electromagnetic radiation is reflected back, either in the green or near infrared, by healthy vegetation not all the incident radiation is returned. The remainder can be accounted for by absorption but also by transmittance through spaces between cells and leaves (Figure 2.13). The transmittance of electromagnetic radiation (as opposed to the reflectance or absorption) also varies with wavelength for vegetation. Within the visible range the transmittance of radiation is low but, like reflectance, it also increases greatly within the near infrared. Infrared radiation that is transmitted through one leaf is then available for reflection, absorption and transmission by underlying leaves. In a forest canopy, this process may extend through six to eight layers of leaves; a thin canopy (two to three layers) would thus be expected to have a lower infrared signature than a thicker one. The degree of transmittance will also vary depending on the type and density of the vegetation canopy. For a densely growing single-species stand, such as a field of wheat, little electromagnetic radiation penetrates beneath the vegetation leaves to the soil, and responses in both the green and the near-infrared parts of the spectrum will therefore be strong. However, in an open forest, with spaces between the trees, radiation of all wavelengths will penetrate to lower levels of understorey vegetation and thus the response from the surface of the top layer will not be as strong. Some of the rays which penetrate the canopy will be reflected back but may be travelling in a direction such that they will be intercepted by the underside of the upper layer of the canopy and reflected back down towards the ground. This phenomenon is termed 'internal reflectance' and has the effect of reducing the total amount of electromagnetic radiation returning to the sensor. Thus, if you view a mixed rural landscape of healthy green vegetation, woodland areas may appear a darker green than pasture lands. The same effect occurs within the near infrared. The response of vegetation is much more complex within the microwave part of the spectrum as the wavelengths of the electromagnetic radiation are a size comparable with the components of the vegetation canopy (branches, leaves, seed pods). This causes different types and degrees of scattering.

and 1.9 µm are due to the presence of water within the vegetation which absorbs radiation at these wavelengths. It must be re-emphasised that Figure 2.11 illustrates a generalised representation of healthy green vegetation but it allows us to develop a number of important points.

First, if similar graphs are produced for a range of vegetation types, they are often quite similar in the visible range but may have a markedly different reflectance in the near infrared. This is shown diagrammatically as the discontinuous line in Figure 2.11. For example, deciduous trees have reflectivity characteristics similar to those of coniferous trees in the visible range but they have a much higher reflectance in the near infrared. It is possible to be more species-specific: for example, birch has a higher reflectance than fir trees. This wide range of signatures for different species within the infrared makes this part of the electromagnetic spectrum particularly important for vegetational discrimination.

Second, we are able to ascertain visibly when vegetation dies by a change in its colour. This is illustrated most dramatically in autumn, when green leaves turn to a yellow or red/brown colour. Similar changes occur within the infrared, which we cannot detect with our eyes. However, vegetation may become stressed by lack of water or a parasite infestation and changes in the signature within the infrared occur much earlier than changes within the visible range become apparent. Consequently, if vegetation is monitored within the infrared, problems that are developing may be recognised and early remedial action taken before they even become apparent in the visible range.

The red edge does not occur at a specific wavelength for all types of vegetation. One might assume from this that measuring the wavelength at which the red edge occurs would indicate the vegetation

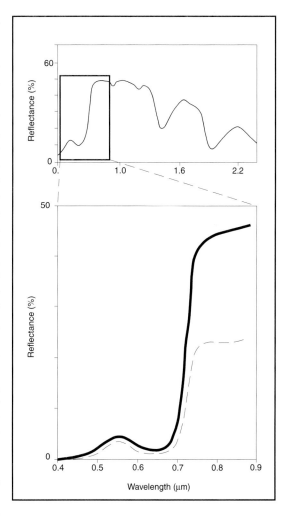

Figure 2.11 Reflectivity curve for green vegetation in the visible and near infrared. The major characteristic of this curve is the marked increase in reflectance at the boundary of the red and infrared (approximately 0.7 µm). The discontinuous line in the bottom diagram shows the curve for stressed vegetation. Note how the distinction between healthy and stressed vegetation can be much more easily determined in the infrared than in the visible range.

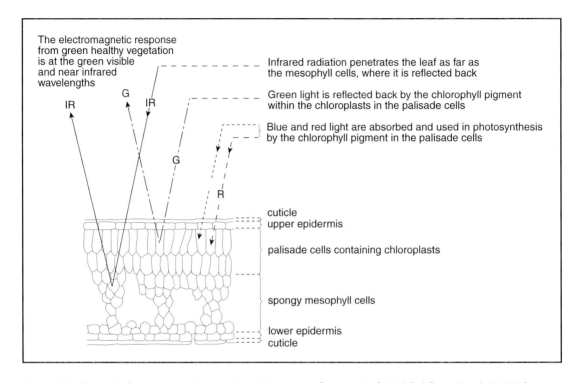

The electromagnetic response from green healthy vegetation is at the green visible and near infrared wavelengths

IR   G   IR

G

R

Infrared radiation penetrates the leaf as far as the mesophyll cells, where it is reflected back

Green light is reflected back by the chlorophyll pigment within the chloroplasts in the palisade cells

Blue and red light are absorbed and used in photosynthesis by the chlorophyll pigment in the palisade cells

cuticle
upper epidermis

palisade cells containing chloroplasts

spongy mesophyll cells

lower epidermis
cuticle

Figure 2.12 Generalised cross-section showing the cell structure of a green leaf. Modified from Campbell (1996).

species. To achieve this, very narrow wavebands would need to be measured; a further complication is that the red edge itself shifts for a single species as the plant grows and matures.

The spectral signature obtained by the sensors on a remote sensing system is often a combination of vegetation and soil. In such a situation, the proportion of soil to vegetation cover will greatly influence the resultant signature. This proportion will vary through the growing season. The characteristics of the underlying rock may make a significant contribution to the observed signature if the soil has formed *in situ* and is not constructed of glacial material which has been moved a long distance. Basalt, which is an iron-rich rock, can weather under tropical conditions to produce a very prominent red signature in the visible range. The presence of moisture reduces the reflectance across all wavelengths. Organic-rich soils tend to have a lower reflectance than organic-poor soils.

Water in the Liquid and Solid State

Water is essential for life on Earth: over 70 per cent of the Earth is covered by liquid water, 10 per cent by ice, and water vapour in the form of clouds also covers large parts of the globe. From a remote sensing perspective, its reflectance characteristics vary significantly depending on whether it is in the liquid or solid phase. The reflectance of water in a solid phase such as in the form of ice or snow is very high at all visible wavelengths whereas the reflectance of liquid water in the visible part of the electromagnetic spectrum is very low. The difference in reflectance properties is not due to chemical differences such as might occur between rocks containing different minerals; only H_2O is present in ice and water. The difference in reflectance is due to differences in how the atoms bond.

The effects of atmospheric water vapour on electromagnetic radiation of different wavelengths have been

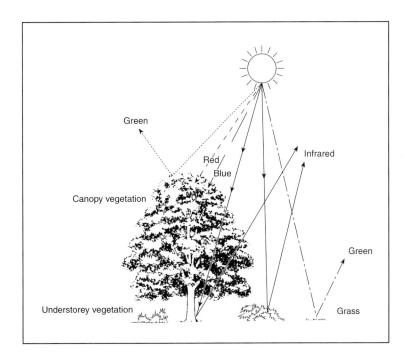

Figure 2.13 How a forest canopy and layered vegetation reflect spectral radiation.

discussed earlier. This section concentrates on terrestrial water bodies: lakes, rivers, ponds, seas and oceans. The spectral response from a water body is complex, as water in any quantity is a medium that is semi-transparent to electromagnetic radiation. It is analogous to the atmosphere and consequently electromagnetic radiation may be absorbed, scattered and transmitted. The spectral response also varies according to wavelength, the nature of the water surface, the optical properties of the water and the angles of illumination and observation of reflected radiation from the surface and the bottom of shallow water bodies.

Pure clear water has a relatively high reflectance in shorter wavelengths between 0.4 and 0.6 μm and virtually no reflectance at wavelengths greater than 0.7 μm. Thus clear water appears black on an infrared image (see, for example, Figure 2.19). However, shorter wavelength reflectance is further complicated by the fact that the maximum transmittance of visible wavelengths occurs between 0.44 and 0.54 μm, also within the blue/green part of the

spectrum. Thus both reflection and transmittance take place at these wavelengths. Shorter wavelengths can penetrate deeper into a water body than longer wavelengths (Figure 2.14a). Once the radiation at these short wavelengths has penetrated the water body, it meets component particles in the water, some of which are of similar size to the wavelength and therefore scattering occurs in much the same way as in the atmosphere (Figure 2.14b). The colour or response of a water body is determined by the radiation which is scattered and reflected within the body itself, not from its surface. This is termed 'volume reflection' as it occurs over a range of depths. In shallow, calm, clear water bodies up to 30 m deep, electromagnetic radiation can penetrate to the bed of the water body. In this case, the colour and nature of the bed material also influence the response for the water.

Water containing a heavy load of sediment, for example estuarine water, is termed 'turbid'. The sediment is suspended within the body of the water and

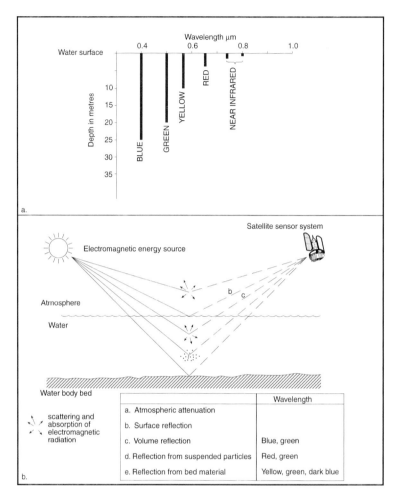

Figure 2.14 (a) Depth of penetration of radiation with different wavelengths into a clear calm water body. (b) The complex spectral response from a water body, illustrating volume reflection. Modified from Campbell (1996).

tends to increase the reflectivity at longer wavelengths of the visible part of the spectrum (the yellow/red range). A reflective response from a water body within the infrared is usually a result of an algal bloom or near-surface weed and/or phytoplankton, i.e. the signature is influenced by a healthy 'vegetation' response. Surface roughness of the water body will also determine how far electromagnetic radiation penetrates it. If the surface is very rough or is viewed from a low angle, then no penetration occurs and the water body response comprises equal reflection of all wavelengths from the surface, producing a bright response, e.g. sun glints (see Figure 2.9). Surface roughness is also important at microwave wavelengths when the irregularities of the surface are similar in length or larger than the wavelength of the incident radiation. In such a situation, the radiation is reflected and scattered back towards the sensor and the water body has a white signature.

Rocks

Rocks are formed of aggregates of minerals. The reflectance characteristics of rocks are thus controlled to some extent by the reflectance properties of their constituent minerals. Reflectance spectra for minerals are obtained by laboratory experiments. Sample minerals are illuminated with light of a known intensity and wavelength and the amount reflected is compared to a standard compound. The percentage reflected is much higher for quartz and feldspar (around 100 per cent) than for olivine and pyroxene (50 per cent). Thus, in the visible range, granite, which contains quartz and feldspar, is paler (i.e. has a higher reflectance) than gabbro, which has pyroxene and olivine minerals.

Pure quartz and feldspar produce featureless spectra: the reflectance is thus approximately constant within the visible and near infrared. Other minerals, however, are characterised by absorption bands at particular wavelengths due to the presence of specific elements within the crystalline lattice of the minerals. The most important element producing a distinctive absorption band in the visible and near infrared is iron, which may be present in olivines, pyroxenes, micas and amphiboles. An absorption band at 0.9–1.0 μm is due to the presence of ferrous iron (Fe). Iron in a more oxidised format (ferric Fe) is associated with an absorption band at 0.7 μm. The presence of water and hydroxyl ions produces characteristic absorption bands in the near-infrared part of the electromagnetic spectrum. Absorption bands at 1.4 μm and 1.9 μm indicate the presence of undisassociated water whereas hydroxyl ions, which are also present in clays, can yield absorption bands at 1.4 μm, 2.2 μm and at 2.3 μm. Carbonates also produce characteristic absorptions at around 2.2 μm. In order for the spectral differences to be distinguished by remote sensing techniques, the diagnostic features must also coincide with an atmospheric window. Figure 2.5b shows the location of atmospheric windows and some water, hydroxyl and carbonates absorption bands fall within or on the edge of such windows. Another factor that has to be borne in mind regarding the use of remote sensing for geological mapping is that, within the visible and near infrared, the sensors obtain their data from the surface of the rock. The mineralogy and the reflectance of a weathered surface are often markedly different from the bulk rock. Two aspects of the weathering surface are of particular relevance regarding remote sensing.

1 In arid terrain, 'desert varnish', which is a thin patina of iron and manganese compounds drawn to the rock surface by capillary action, may coat a range of rock types, reducing their reflectance, and produce a dark spectral response for different rock types. The satellite image shown in Figure 2.15 encompasses a range of rock types, but little distinction can be made between them because of desert varnish.
2 Oxidation of iron-rich rocks can produce very distinctive weathering products that may be detected by remote sensing techniques. Hydrothermal alteration can occur over ore bodies, which can yield minerals which produce distinctive spectra. The ore body itself may be unexposed or too small to be detected by remote sensing but the alteration zone can be considerably larger and may be targeted and thus indirectly indicate the location of mineralisation.

Although different rocks will often appear different on a remote sensing image, based on spectral signatures alone, it is not possible to assign a unique spectral signature to a specific lithology which will be applicable in all circumstances, using sensors operating on most remote sensing systems today within the visible and near infrared.

Cultural Structures

It is not possible unequivocally to attribute specific reflectances to particular cultural features because of the great variety of materials that are often used for the same feature. Thus a road surfaced with tar will have a low reflectance in the visible range and be virtually black in the near infrared whereas a road constructed of concrete sections will have a higher reflectance in all wavelengths. Similarly, buildings may be roofed with natural materials such as slate or artificially produced tiles in a variety of colours.

Characteristic reflectance curves for a range of natural substances within the environment are shown together in Figure 2.16.

Figure 2.15 The effects of desert varnish. Although a range of rock types occurs in this region of Sudan and Egypt (sandstone, schist, granite, serpentine), the presence of desert varnish results in a low, relatively uniform reflectance. Image approximately 110 km wide.

Factors that Affect Remote Sensing Signatures

An ideal remote sensing system would allow us to measure a characteristic of a body, such as its reflectance, at any wavelength that we choose. As mentioned previously, this is not possible because the atmosphere selectively absorbs electromagnetic radiation of particular wavelengths and the signature obtained also contains a scattering component. Other factors can affect the signature of an object to a greater or lesser degree depending upon the actual purpose of the investigation. For example, cloud cover would cause problems to a botanist who is attempting to measure the reflectances for different types of vegetation using spaceborne sensors whereas to a meteorologist the presence of clouds is a distinct advantage. Similarly, a geologist who is interested in mapping rock structures would prefer a total absence of vegetation as it masks the rocks.

Passive remote sensing systems use the Sun as their source of illumination. However, the direction of this illumination changes throughout the day and the illumination angle changes throughout the year (Figure 2.17). In the northern hemisphere the Sun is seen to rise (approximately) in the east and set in the west. Features, especially linear ones, trending at right angles to the illumination direction tend to be highlighted whereas features parallel to the illumination direction often appear more subdued and may go unnoticed during analysis (Figure 2.17a). Images can be obtained on aerial surveys at different times of the day. Thus if the same area is imaged in the morning and late afternoon, a greater number of features with different orientations may be discerned than if the area is only imaged in the morning. However, many satellite systems (such as Landsat and SPOT) are put into an orbit that allows the onboard sensors to acquire their data at approximately the same local time for all parts of the globe, usually in the morning. This has the advantage that different parts of the Earth are imaged under a constant illumination direction, but as a consequence, features that are parallel to this direction (usually NW/SE-trending structures in the northern hemisphere) are suppressed.

A combination of the Earth's rotation, its orbit around the Sun, the tilt of the Earth's axis and its sphericity produces the seasons of winter, spring, autumn and summer. At any given time the apparent position of the Sun in the sky will be lower in the winter than in the summer. The low winter illumina-

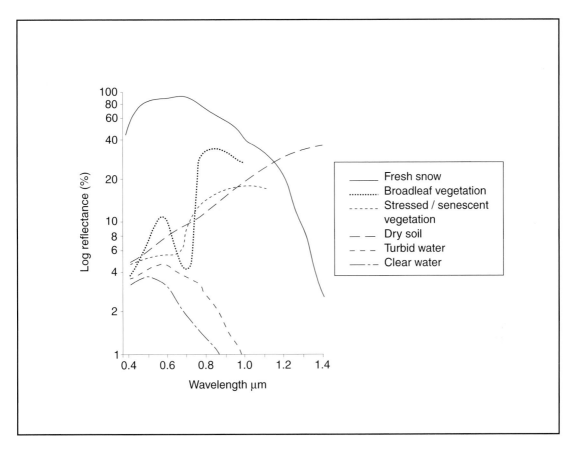

Figure 2.16 The spectral response (reflectance) curves for different natural surfaces at different wavelengths.

tion angle accentuates topographic features because of the much longer shadows they cast (Figure 2.17b). Thus a feature might not be detected on a high illumination angle summer image but can be prominent on a low-angle winter one. The yearly variation in angles changes with latitude. At the equator there is little difference in the illumination angle throughout the year whereas, at higher latitudes, a summer image may have an illumination angle 50 degrees greater than a winter image.

Satellite images of northeast Ireland obtained of the same area at the same time of the day in April and December are shown in Figure 2.18. When the April image was obtained the Sun was relatively high in the sky with an illumination angle of 35 degrees. Little topographic information can be obtained from this high solar angle. However, on the image obtained in December (Figure 2.18b) when the solar illumination is 10 degrees, NE/SW-trending sinuous ridges can be observed which are part of the North Antrim End Moraine which formed during the last glacial period. A distinct textural change can also be discerned across the moraine on the low-angle illumination image which is not observed at the higher illumination angle. Extensive drumlin fields south of the moraine, which are absent to the north, produce a mottled 'hummocky' signature. Although topographic differences are best observed at low illumination angles, tonal differences are generally more noticeable at higher illumination angles.

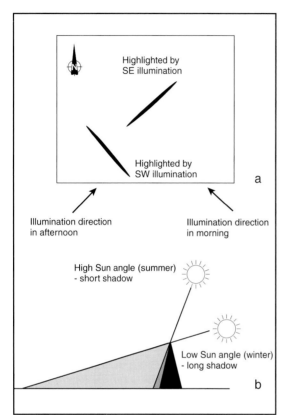

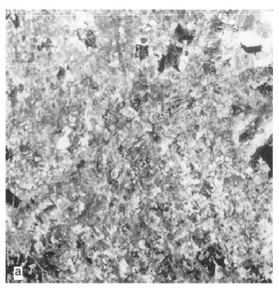

Figure 2.17 Effects of solar elevation and direction on the detection of features on remotely sensed images. (a) Features trending parallel to the illumination direction tend to be subdued whereas features at right angles are highlighted. (b) Low-angle illumination highlights topographic features whereas a higher Sun angle allows a better discrimination of tonal variations.

Figure 2.18 (a) High-angle illumination Landsat TM image of northeast Ireland allows field patterns to be determined but there is little topographic information. (b) The same region viewed at a low angle clearly displays the sinuous ridges of the Armoy End Moraine. 15 x 15 Km.

2.4 TYPES OF REMOTE SENSING IMAGE

It is not possible to devise a single image-classification system which will be the optimum one in all circumstances. The term 'recording' was used in the definition of remote sensing given in Chapter 1 and the classification presented here is based on the two methods by which the remotely sensed data are recorded: photographic and digital.

Photographic Remote Sensing

In photographic remote sensing, the representation of a scene is recorded by a camera onto photographic film. Electromagnetic radiation passes through the lens at the front of the camera and is focused on the recording medium (film). The characteristics of the film (and also the lens and the nature of any filters that are being used) determine the signal that is recorded. Different types of remote sensing photographs may be produced, each of which has advantages and disadvantages. The main types are:

1 Panchromatic
2 Photographic infrared
3 Multispectral
4 Natural colour
5 False colour.

Panchromatic Photographs

The simplest film is one that records variations in electromagnetic radiation within the visible range of the spectrum (0.4–0.7 µm) in black and white and shades of grey. The resultant image is a panchromatic photograph but is often referred to as a black and white photograph. The acquisition of a panchromatic photograph is a multi-stage process. When the shutter of a camera is opened, the film is exposed to light reflected from a surface. The film is coated with compounds that are sensitive to this light which produces a chemical change in their nature. The degree of change is proportional to the amount of light that reaches the film: bright parts of the scene will produce a marked change in the composition of the film whereas very dark areas of the scene, i.e. those from which the light intensities are very low, will produce virtually no change in the film. Although a representation of the scene is now recorded on the film, it cannot be observed (often referred to as a latent image) until further chemical treatment of the film has taken place. This treatment results in the production of a negative, in which tonal relationships are reversed: bright parts of the scene are represented by dark portions of the negative and vice versa. In order to obtain a positive print, i.e. one in which the tonal relationships are direct, it is necessary to project light through the negative on to a sensitised medium. Transparent areas of the negative (which represent dark parts of the scene) will allow a large amount of light to reach the sensitised paper, which, after suitable processing, will appear dark – similar to the original scene. A panchromatic photograph has a number of advantages over other types of remotely sensed image but there are also disadvantages.

Panchromatic film production technology has been well developed over the years and panchromatic films are relatively cheap. In addition, the film does not require sophisticated processing. Another advantage of panchromatic aerial photography is that most countries have extensive coverage of this type of image extending back over many decades. Having photographs recorded on the same medium allows comparisons to be made easily and any changes to the landscape can be quickly identified. A disadvantage of panchromatic film is that, while many features can be identified on black and white photographs such as roads, bridges and buildings, many other objects are not so distinguishable. Differentiating between different surfaces such as different crops is often impossible. In addition, the tone of any particular feature is due to the combined effects of the reflectances across the visible range. Thus two features may have an identical tone but the tones may have been caused by different spectral reflectances. One body may have a high reflectance in the blue and green and low in the red, whereas the other may have a low reflectance in the blue but a high one in the green and red. Examples of panchromatic aerial photographs are shown in Figure 4.14.

Black and White Infrared Photography

A typical black and white infrared film that is used in aerial photography is the Kodak Infrared Aerographic Film 2424. The sensitivity of this film, unlike a panchromatic one, extends into the near infrared. However, this infrared film has a greater sensitivity in the ultraviolet range than in the infrared part of the spectrum. The lower wavelengths may be filtered out (see next section), which effectively makes the film most sensitive to a waveband extending from 0.76 μm to 0.88 μm.

Some of the advantages of infrared film are:

1 The use of this film extends the range at which the spectral reflectivity characteristics of different objects can be examined.
2 Scattering on infrared photographs is less than on panchromatic photographs because of the longer wavelength at which infrared operates. Thus they often appear sharper.
3 Infrared images are ideal for delineating land–water interfaces because infrared does not penetrate water. The contact may be sharply defined because water has a black signature whereas the land, especially if it is vegetated, may appear much brighter. Satellite photographic infrared images are particularly useful in the search for water in arid regions. Dried-up wadis yield a bright signature in the infrared because of the presence of highly reflective sand which contrasts sharply with the black signature associated with water. Thus a satellite image which may cover tens of thousands of square kilometres can be quickly examined and drainage systems which contain water can easily be distinguished from those that are barren (Figure 2.19).
4 Infrared images are ideal for vegetation surveys. Signatures for an ivy-clad building fronted by a grass lawn illustrate the reflectance of vegetation in the photographic infrared compared with the visible range (Figure 2.20). The reflectance for the grass and ivy is substantially greater in the infrared. The distinction between the low-reflectance path and the high-reflectance grass in the foreground on infrared allows these two surfaces to be easily differentiated but in the panchromatic photograph the distinction is much less obvious. The poplar and beech trees behind the building cannot be differentiated in panchromatic whereas on the infrared photograph the poplar tree is seen to have a higher reflectance than the beech.

Black and white infrared images also have some disadvantages:

1 There is a lack of detail in shadowed areas resulting from the very low scattering. A cylindrical

Filters

An alternative means of viewing the role of complementary colours is in the production of filters. A filter is a gelatine or glass cap which fits over the camera lens and prevents electromagnetic radiation of specific wavelengths being transmitted by the lens and reaching the film. Filters are employed in black and white and colour photography. The primary additive colours are transmitted by filters that are the same colour, i.e. a red filter transmits red light but blocks green and blue. A complementary colour blocks its complement but transmits the other two additive primary colours. Thus a yellow filter blocks blue light but transmits green and red. This is particularly useful for reducing haze in an image. Scattering affects shorter wavelengths to a greater extent than longer wavelengths, so that removing the blue component yields a clearer image. A panchromatic photograph obtained by using this type of filter is referred to as a minus-blue photograph.

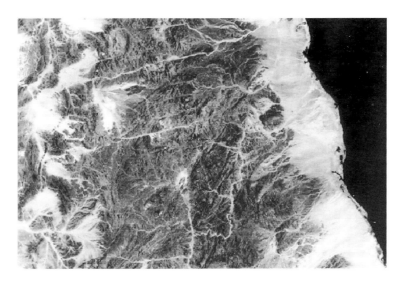

Figure 2.19 Infrared image of part of the Red Sea Hills, Sudan. The very high reflectivity associated with the linear wadis shows that no water is present at the surface. Compare this signature with the Red Sea in the east. Width of image is 135 km.

tank to the side of the building on Figure 2.20 is obvious on the panchromatic image because it is indirectly illuminated by scattered radiation. However, scattering at the longer infrared wavelength is minimal, little information can therefore be obtained in the shadows and the tank is not detected.

2 The lack of penetration of water means that objects just below the surface will not be detected. Thus sandbanks which may present a danger to shipping will not register on an infrared image. Panchromatic and infrared photographs of a river in Britain are illustrated in Figure 2.21. The panchromatic image shows the presence of a small pale sand bar. However, this feature is actually below the water surface, which is not penetrated by the infrared.

Natural Colour Photography

Although black and white images can often provide information about a particular area, in order to differentiate surfaces it is often necessary to employ colour images. Different colours can be produced by an additive combination of red, green and blue light which are known as the primary additive colours. If the primary colours are superimposed in combinations of two, three other colours can be produced which are known as complementary colours. Thus the superimposition of red and green light of equal intensity produces yellow; green and blue produces cyan and red and blue yields magenta (Plate 2.1a). The complementary is obtained by mixing the two other additive primary colours. Thus yellow is the complement of blue; magenta is the complement of green and cyan is the complement of red. When the three additive primary colours are superimposed in unequal amounts, a range of colours is produced. If they are superimposed in equal amounts then greys, ranging from black to white, are formed.

The advantages of normal colour photographs are:

1 Surfaces that are indistinguishable on a black and white image can often be easily differentiated on a colour image because the human visual system can differentiate hundreds of thousands of colours but relatively few grey levels.

2 The colours produced accord with our perception of everyday living. They are thus relatively easy to interpret.

Figure 2.20 Reflectivity of vegetation in the (a) visible and (b) infrared parts of the electromagnetic spectrum. Note also the lack of information that can be obtained within the shadows on the infrared photograph because there is so little scattering at this wavelength.

3 Much of the expense of an aerial survey involves the cost of hiring the aircraft and personnel. These are constant whether colour or black and white images are obtained.

The disadvantages of colour photography are:

1 More complex and expensive processing of the film is required.

2 Colour images can have less definition than black and white images.

3 The colours may disrupt the continuity of linear features which cross differently coloured surfaces.

4 The colour photograph may contain too much 'distracting' information.

Figure 2.21 Comparison of aerial photographs obtained in (a) blue and (b) near infrared ranges. The presence of a small sand bar (shown pale) beneath the water can be determined on the photograph obtained in the blue range but not on the infrared photograph. Courtesy and © Hunting Technical Services and Hunting Aerofilms.

Multispectral Images

Multispectral photography involves simultaneously obtaining images of the same scene at different wavelengths. The most common arrangement for multispectral imaging is the acquisition of four images in the blue, green, red and photographic infrared parts of the spectrum. An example is shown in Figure 2.22 which illustrates that surfaces can often be differentiated better at particular wavelengths. The road has a very high reflectivity in the blue part of the spectrum which contrasts well with the lower reflectance surroundings and the very dark signature for the river. The road is not so prominent on either the green or

red image though it is sharply delineated on infrared where it has a low reflectance. However, in the infrared both the road and the river have the same low reflectance. Deciduous trees on the infrared image are seen to have a much brighter signature than on the other bands. The signature obtained for the industrial plant (bottom edge of images) also varies considerably with wavelength, much more information being obtained in the green range (Figure 2.22b) compared with the red (Figure 2.22c).

Multispectral imaging allows the examination of single-band images; natural colour and false colour images can also be produced. A question that may legitimately be asked is why we should produce false colour images (also referred to as 'false colour composites') when we are able to produce normal colour ones. Consider the hypothetical situation shown in Figure 2.23, where the reflectances for two surfaces (shown diagramatically by the length of the bars) in the blue, green, red and photographic infrared have been obtained. If the situation shown in Figure 2.23a prevails, where the reflectance for surface A is different from surface B in all wavelengths except for the infrared, it would not be appropriate to produce a false colour image as a normal colour image will best display the differences between them. Thus the reflectances measured in the blue, green and red ranges are projected in blue, green and red light and superimposed in order to produce the natural colour image. However, if the measured reflectances are as shown in Figure 2.23b, then a normal colour image is not the best option. In this instance, as the reflectance obtained in the blue range is identical for both surfaces, including the blue range, information in an image does not contribute to the differentiation of the surfaces. It makes much more sense to produce an image by using the green, red, and infrared ranges where the two surfaces have different reflectances. This presents us with a problem, as we cannot see infrared. However, projecting the green range in blue, the red range in green and the infrared range in red light can produce a false colour image. Note that, because the false colour is produced by consistently projecting in light of a shorter wavelength than the wavelength at which the data were obtained, the

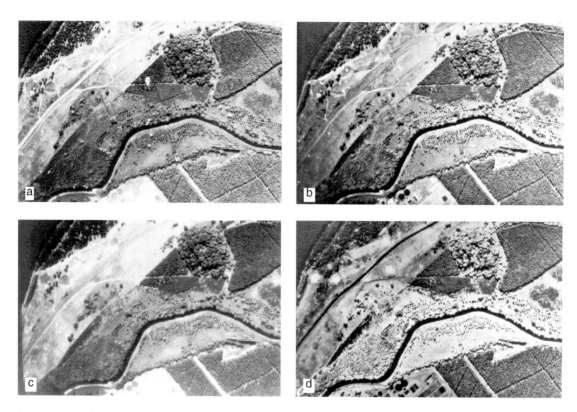

Figure 2.22 Multispectral photographs in the (a) blue (top left); (b) green (top right); (c) red (bottom left) and (d) infrared (bottom right) ranges. Courtesy and © Hunting Technical Services and Hunting Aerofilms.

colour of a feature on the false colour composite is at a shorter wavelength than the feature would appear in reality.

A false colour composite, which is formed by projecting green data in blue, red in green and infrared in red, is known as a standard false colour composite. Many remote sensing systems obtain data in a number of bands at longer wavelengths than visible light. Landsat TM, for example, records seven bands, four of which cannot be detected by the human visual system (see Table 3.1). Thirty-five different colour images can be produced using the seven Landsat TM bands. (Note, the Landsat bands data are acquired digitally, and this will be discussed later. However, the theory behind the production of false colour composites is similar.)

A major advantage of multispectral imaging is that a degree of flexibility is introduced to the data. The scene may initially be examined separately in the blue, green, red and infrared parts of the spectrum and as a natural or false colour composite in which information in parts of the electromagnetic spectrum which is invisible to the human eye can be displayed on an image. A disadvantage is that the camera system is much more complex than that for obtaining panchromatic or natural colour photographs and exposure times for the various films have to be matched. The production of colour images from the individual black and white ones is also more complicated than if a single colour film is used.

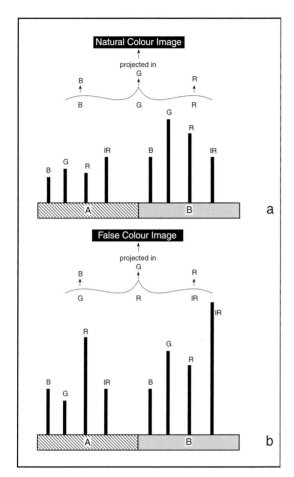

Figure 2.23 (a) The formation of a natural colour image by projecting in blue light the data collected in the blue range of the spectrum, projecting in green light the data collected in the green range of the spectrum and projecting in red light the data collected in the red range of the spectrum. (b) The formation of a false colour image by projecting in blue light the data collected in the green range of the spectrum, projecting in green light the data collected in the red range of the spectrum and projecting in red light the data collected in the infrared range of the spectrum.

Example: What colour would represent green vegetation on a false colour composite?

If you answered 'blue' to this question, your logic is impeccable but your answer is wrong! This is because you neglected to take account of the very high reflectance of vegetation in the infrared. As Figure 2.11 shows, the false colour composite for vegetation is formed by a little blue (due to reflectance in green), virtually no green (due to very poor reflectance in red) and a very large red component (due to extremely high reflectance in the infrared). Thus vegetation will appear red on a false-colour composite. Water, because of a low reflectance in green, red and infrared wavelengths, will appear black. Blue objects, if they have a low reflectance in the infrared, will also appear black on a false colour composite. The vegetation on the aerial infrared image shown in Plate 2.2 shows this characteristic red colour. The healthy deciduous trees along the main street of the town stand out very clearly.

Digital Images

Although images can be produced using film as the recording medium, it is also possible to record a scene electronically and form an image of it. Digital images are of great importance in remote sensing for a number of reasons.

1 Film cannot record electromagnetic radiation at wavelengths longer than about 1 μm and consequently no data can be obtained in the thermal and microwave bands by camera systems. Thus, if we wish to exploit information at these wavelengths, a different type of recording system is required.

2 Before 1960 the film obtained on aerial surveys could be recovered and processed when the aircraft landed. However, satellite systems since 1960 operate at very high orbits and are designed to produce large amounts of data, often for years. It

Subtractive Colour Printing

Although various colours can be produced by an additive process involving the combination of red, green and blue in different amounts, colour printing, such as the production of the colour images shown in this book, is effected by a subtractive process. If white light is passed through a coloured filter, it is possible to subtract one of the primary colours (red, green or blue) and pass through the remaining two colours (Plate 2.1b). Thus a yellow filter subtracts blue light but passes green and red; a magenta filter subtracts green light and passes blue and red and a cyan filter subtracts red light and passes green and blue.

Yellow, magenta and cyan are known as subtractive primary colours. If two filters are overlapped, only one colour passes through. Yellow and magenta filters allow red through, yellow and cyan allow green to pass and magenta and cyan allow blue light to pass. No light may pass through a sandwich of all three filters and a black colour is registered. Colour film consists essentially of a sandwich of three emulsions, each of which is blue, green or red sensitive because every colour observed in a scene can be thought of as formed of these additive primary colours. Upon exposure, variations in the blue component of the scene are registered on the topmost layer which is blue sensitive, green variations in the middle layer, which is green sensitive, and red variations on the bottom layer, which is red sensitive. In a fashion similar to black and white photographs, the latent images preserved on the three emulsions have to be developed in order to make them visible. Developing the film converts all the exposed silver halide to black but in the process colour couplers result in the formation of dyes on each of the emulsions. Bleaching removes the black silver grains producing a colour negative. However, this negative shows blue, green and red in the original scene by yellow, magenta and cyan dyes respectively. An inverse relationship exists between the amount of additive colour in the scene and the amount of dye in the negative. Consider a very simple scene consisting of a pure red flower, growing in pure green grass with a pure blue sky behind. On the negative the red flower will contain no cyan component (because of the inverse relationship) and will be represented by magenta and yellow. When white light is shone through the negative, the magenta and yellow combined will only allow the passage of red light through the flower, i.e its original colour. The green grass on the negative will be formed of cyan and yellow and no magenta; cyan and yellow combined allow the passage of green light. The blue sky will contain no yellow on the negative but is formed of cyan and magenta, which together allow only green light through.

Various diagrams exist which seek to relate the various colours and their roles. The simplest is one which consists of two overlapping triangles, one with an additive colour at each apex and the other with a subtractive colour at each apex (Figure 2.24). Colours on any side of a triangle make up the colour shown between them on the other triangle.

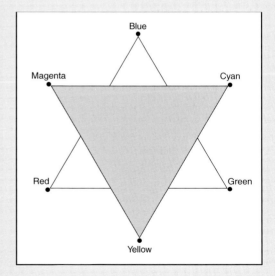

Figure 2.24 Star diagram illustrating relationships between the additive and subtractive primary colours.

Thus green and blue additive colours produce cyan while cyan and magenta subtractive colours allow blue light to pass. The additive on an apex does not contribute to the colour opposite it on the subtractive triangle. Thus blue does not contribute to the formation of yellow, which is consequently the complement of blue. If one wants to make an image more blue this can be achieved by either adding more magenta and cyan or subtracting more yellow.

is neither feasible to recover film from these satellites nor practical to launch an expensive satellite, which because of weight restrictions could carry only a limited supply of film. (Exceptions to this rule are military reconnaissance satellites, where expense is not the overriding factor.)

3   Solid-state electronic devices are very reliable, use little power and are small and light. All these properties are extremely important for satellite systems. Once the satellite is launched (apart from very low-orbiting ones) it is not possible to replace the components. In addition, weight is always at a premium in satellite systems.

4   Data that are obtained digitally can be transmitted easily without any degradation.

5   Digital data are in a form that can be readily processed on computers. Digital image processing techniques are extremely powerful algorithms that allow maximum information extraction from the data.

A digital image is a regular grid array of squares (or rectangles) where each square is assigned a number which is related to some parameter (such as reflectance or emittance) which is being measured by a remote sensing system's sensor. Figure 2.25 illustrates a simple scene which is characterised by a range of reflectances from very dark to very pale. The equivalent digital image, which is a representation of this scene, is shown below it. As can be seen, the very dark areas are associated with very low digital numbers (DNs) and brighter parts with high digital numbers. By convention the rows and columns are numbered with the origin at the top left, so that the small area represented by the number 26 is row 1, column 1 or more simply (1, 1). Each square that is assigned a DN is referred to as a 'pixel', which is a word formed from the term 'picture

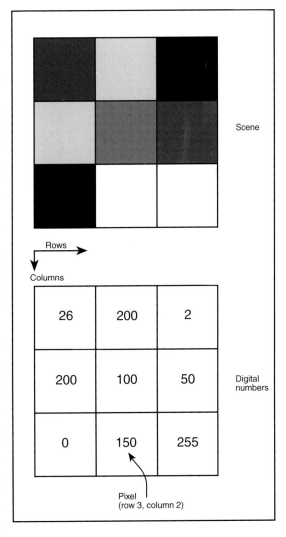

Figure 2.25 Simple representation of a digital image in which the scene is divided into a regular array of pixels. Each pixel is associated with a digital number which can be related to some property (such as reflectance) of the scene.

element'. There are two main system types employed by passive remote sensing systems for the acquisition of digital data; transverse scanning systems and pushbroom systems.

Transverse Scanning System

A transverse (also termed across-track or whiskbroom) scanning system is an electro-mechanical device that obtains data from narrow swaths of terrain, which are at right angles to the direction of movement. A scanning mirror sweeps across the scene and then directs this reflected (or emitted) radiation towards the onboard detectors (Figure 2.26a). The continuous signal generated by the mirror movement is split into a number of wavebands depending on the spectral resolution of the system (in the example shown in Figure 2.26a there are three bands and hence three detectors) and sampled at regular short time intervals. The short interval represents a specific sweep length on the ground and is referred to as a sampling cell. Depending on the strength of the signal obtained for this sampling cell, a DN is allocated to it. The forward movement of the remote sensing platform (aircraft or satellite) allows

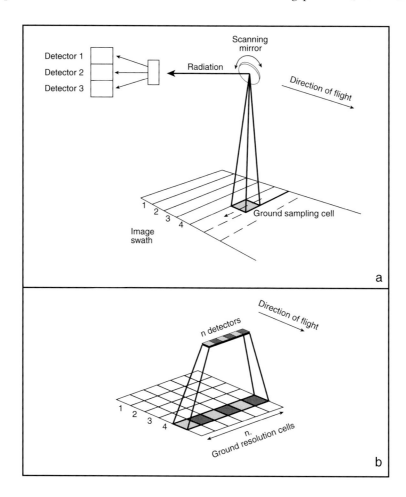

Figure 2.26 Two main types of system for obtaining digital data: (a) transverse scanning systems (also known as across-track) which employ an oscillating mirror and (b) pushbroom systems which employ a detector for every pixel in a line.

the next line of data to be obtained (in the order 1, 2, 3, 4 on Figure 2.26a), and thus the image is built up in a sequential fashion, unlike an aerial photograph, which obtains its data for an entire scene virtually instantaneously. Most transverse scanning systems obtain their data from directly beneath the platform and to either side of it. The width of the image that is produced depends on the angle through which the mirror rotates and the height of the platform. A satellite system can image the same swath width as an airborne system using a smaller scan angle which reduces distortions. The scanning mirror traverses the scene with a constant angular velocity. Thus, in time (t) the ground distance scanned (y, Figure 2.27) below the platform (nadir) is less than the distance scanned off-nadir (x, Figure 2.27). However, these data are recorded at constant rates and this results in scale compression at off-nadir viewing. The off-nadir distortion is reduced when a satellite system is used instead of an airborne on the same swath width.

Pushbroom System

A pushbroom system does not rely on a scanning mirror to direct radiation onto a detector but instead employs a linear array of detectors, in which each detector measures the radiation reflected from a small area on the ground known as a ground resolution cell (Figure 2.26b). Once again, the forward movement of the platform allows the image to be built up. The detectors are extremely small, with dimensions of about 10^{25} m, and are composed of charge-coupled devices (CCDs).

The lack of a moving mirror for a pushbroom system can make it more reliable than a transverse scanner and

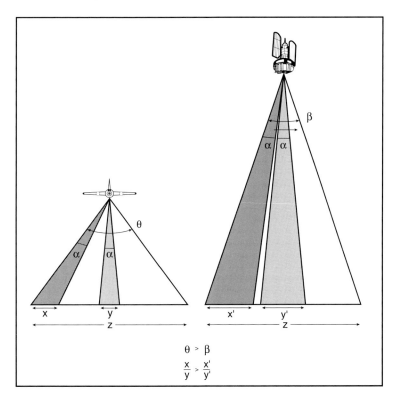

Figure 2.27 Variation in width of area imaged with height of the remote sensing platform and the angle through which the mirror rotates. The width z can be imaged with a smaller scan angle (β) for the satellite than for the aeroplane (θ). The scale distortions across the image are less pronounced for the satellite system.

Computer Storage

One bit in computer memory (which is a shorted version of 'binary digit') can have a value of 0 or 1. In order to represent the values 0–255 (i.e. 256 grey levels), 8 bits of memory are used because in binary format 256 is equivalent to 100000000; 8 bits are referred to as a byte, and thus for image processing systems which use 256 levels, 1 pixel for 1 band is held by 1 byte of memory. A kilobyte is not 1,000 bytes but 2^{10} bytes which equals 1,024. Similarly, a megabyte is 1,024 kilobytes.

the detectors are light and require little power to operate. However, in order for the readings obtained from different detectors to be consistent, they must first be matched and calibrated. Matching the response of so many detectors may be difficult and during their operating lifetime they may degrade at different rates.

Example: A pushbroom system has a pixel size of 10 m and images a 50 km wide swath. How many detectors does the system have and what is the length of the detector array?
Pushbroom systems have an individual detector per pixel. Thus, for a pixel size of 10 m and an image swath of 50 km, 5,000 detectors would be required. If each detector is 10^{-5} m across then this array will have a length of $5,000 \times 10^{-5}$ m or 5 cm.

Resolution

From the brief description given earlier, it can be seen that the concept of a digital image is relatively simple. However, Figure 2.25 hides a number of complexities and a brief examination of it prompts a number of questions. For example, what determines the size of a pixel and what effect does the size of the pixel have on the image? The highest digital number on Figure 2.25 is 255: why not 511 or 1,023, and what difference does it make? Many of these questions can be answered by discussing the concept of resolution. Ask an average person to explain what the resolution of an image is and he/she will most likely say something like 'the smallest distance between two features, so that the two features can still be distinguished from each other'. However, in remote sensing we can consider four types of resolution:

1 spectral
2 temporal
3 radiometric
4 spatial.

Example: How many grey levels are recorded by a 7-bit remote sensing system and what is the DN value for pure white?
A 7-bit system will record 2^7 or 128 grey levels and the highest number recorded is 127. (Remember that the first DN is zero.)

Example: A remote sensing system with a radiometric resolution of 6 bits assigns a DN of 27 to one surface and 45 to another. What would the equivalent digital numbers be for the same surfaces if the measurements were taken with a 3-bit system?
The DNs recorded by the 3-bit system range from 0 to 7 and this range is equivalent to 0–63 for the 6-bit system. Thus:

0	1	2	3	4	5	6	7	(3-bit)
0	9	18	27	36	45	54	63	(6-bit)

Therefore a DN of 27 and 45 on the 6-bit system will be recorded by 3 and 5 respectively on a 3-bit system. Note that a value of 29 on the 6-bit system will also be recorded as 3 on the 3-bit system because the data are compressed.

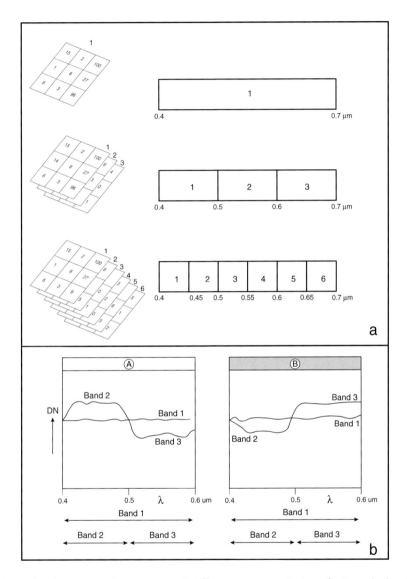

Figure 2.28 (a) Examples of remote sensing systems with different spectral resolutions. (b) Example showing how two surfaces (A and B) are indistinguisable on a single-band image (band 1), but can be differentiated on either band 2 or 3 on a multiband system.

Spectral Resolution

The digital image shown in Figure 2.25 represents the data that have been obtained within a single waveband by a remote sensing system. However, few passive remote sensing systems, especially those operating in orbit, obtain data solely at one waveband because it is not economical to launch a satellite which is restricted to gathering only one set of data. Most remote sensing systems are multispectral and obtain data at a number of wavebands (Figure 2.28a). A multispectral dataset can be envisaged as consisting of a number of stacked layers in which each pixel is associated with a digital number for each layer. Figure 2.28a shows examples of systems that obtain data in one, three and six bands. Each layer of data is referred to as a band (or channel): thus, for example, Landsat MSS obtains data in four bands, and each pixel consequently has four digital numbers associated with it, whereas the Landsat TM system is a seven-band system. The Landsat system is considered in detail in section 3.2. A system which measures a large number of bands which encompass narrow ranges of the electromagnetic spectrum is said to have a high spectral resolution. Thus Landsat TM has a better spectral resolution than Landsat MSS. Some airborne systems measure over 200 bands. Such hyperspectral systems are discussed in Chapter 3. Generally, surfaces can be better distinguished on multiband data sets than on single-band ones. Consider the situation shown in Figure 2.28b, where two surfaces, A and B, are examined at different spectral resolutions. On a single-band image (band 1) the two surfaces are indistinguishable from each other as they have very similar DNs across this broad waveband. However, the two surfaces can be easily distinguished on either band 2 or band 3. Surface A has a higher DN than surface B on band 2 and will thus appear brighter on band 2. The situation is reversed for band 3, where surface A will appear darker.

Temporal Resolution

The temporal resolution of a remote sensing system is a measure of how often data are obtained for the same area. It is generally not applicable to aerial photographic surveys which tend to take place at infrequent intervals, usually for specific projects. However, satellite systems are launched into predetermined orbits at specific altitudes and in these instances it is possible to determine how regularly an area is imaged. The temporal resolution varies from less than one hour for some systems to approximately 20 days for others. Early generations of satellites had a fixed temporal resolution, though later ones such as SPOT (see section 3.2) have onboard off-nadir viewing capabilities to alter the temporal resolution. Some military reconnaissance satellites, although launched into predetermined orbits, do not continually monitor the globe and do not have a fixed temporal resolution. They are often only activated to obtain data when particular 'targets of opportunity' are presented such as over military installations or when the movements of forces by hostile regimes are to be monitored.

Radiometric Resolution

The radiometric resolution of a remote sensing system is a measure of how many grey levels are measured between pure black (which could represent no reflectance from the surface) and pure white. The radiometric resolution is measured in 'bits'. A 1-bit system ($2^1 = 2$) measures only two grey levels and is the simplest type of image. It records black and white (Figure 2.29). An 8-bit system ($2^8 = 256$) records 256 grey levels, in which by convention black is recorded by a digital number of zero and white by a digital number of 255. The higher the radiometric resolution, the greater the number of recorded grey levels for the scene. Most remote sensing systems have a radiometric resolution of 6 bits or higher, though the human visual system cannot detect more than about 30 grey levels.

It is important to realise that, unless the radiometric resolution for all the recorded bands is similar, it is not possible to compare directly the DN across the bands for any pixel. For example, consider a system that obtains data in two bands, band 1 with a 3-bit radiometric resolution and band 2 at 6 bits. A pixel with a DN of 7 on band 1 would be pure white, indicating a very high reflectance at that particular waveband, but the same DN recorded on band 2 (6-bit) would indicate a very low reflectance at that wavelength and be represented by a dark grey signature on a band 2 image.

Spatial Resolution

While radiometric resolution, discussed above, is applicable to digital images, spatial resolution applies to images obtained digitally and by photographic means. No single definition for spatial resolution exists and various techniques have been devised in order that images produced by different

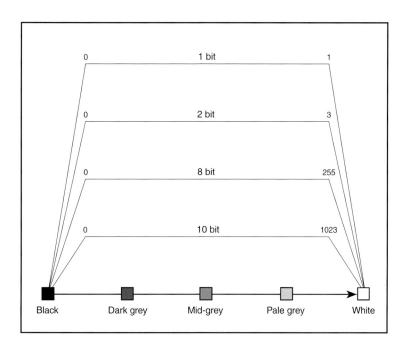

Figure 2.29 Radiometric resolution for a 1-, 2-, 8- and 10-bit system. The greater the number of bits, the greater the number of grey levels that are measured.

systems can be compared and their ability to reproduce a scene accurately can be evaluated. A qualitative measure of spatial resolution is the amount of detail that can be observed on an image. Thus for two images at the same scale and of the same area, the one that shows finer detail may be said to have a better spatial resolution than the one that shows only coarse detail. Distortions due to lens defects can degrade a recorded image and consequently affect the spatial resolution. The quality of a lens can be measured by means of test charts or by producing a modulation transfer function (MTF).

Various types of test charts exist (such as Siemens or Cobb), though a common design involves an array of vertical and horizontal bars of decreasing size (Figure 2.30a). The spacing between the bars is equal to the bar width and the length of the bars is commonly five times the width. A black bar plus its equivalent white space is referred to as a line pair, the reciprocal of which is the spatial frequency. A bar of 0.25 mm is equivalent to a target spatial frequency of 4 which, if

it is photographed with a 10-fold reduction, is equal to an image spatial frequency of 40 line pairs per millimetre (lp/mm). The chart is photographed and examined by means of a microscope at a magnification of 25 in order to ascertain the pattern, which is just distinguishable, i.e. horizontal bars can be distinguished from vertical ones and the bars are observed as separate entities. This effect can be demonstrated by having someone hold the book at eye-level while you move further away. At a particular distance the smallest bars on Figure 2.30 cannot be differentiated whereas the larger ones will remain distinct. Note how it becomes more difficult to resolve the vertical black bars at the bottom of the figure as the thickness and spacing decreases. The chart in Figure 2.30a shows black bars against a white background, i.e. there is a high contrast between the object and background. If the same procedure were performed for dark grey bars against a pale grey background (a low-contrast chart), bars that are just distinguishable at a high contrast would be indistinguishable at a low contrast.

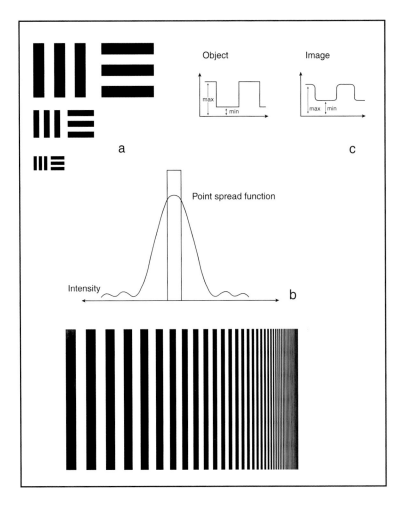

Figure 2.30 (a) Typical text chart for determining spatial resolution. (b) Point spread function. (c) Object and image intensity modulation.

A digital imaging system also incorporates lenses in order to focus the incoming radiation onto the detectors; the optical qualities of these lenses will thus have a bearing on the spatial resolution. However, a digital image is composed of discrete pixels, the size of which sets a lower limit on the spatial resolution that can be achieved. A measure of the size of the pixel is given by the Instantaneous Field Of View (IFOV), which is dependent on the altitude and the viewing angle of the sensor (Figure 2.31). A narrow viewing angle or a lower altitude produces a small IFOV. For a pushbroom system the number of detectors influences the spatial resolution. A system with 1,000 detectors that images a 50 km wide swath has a pixel size of 50 m whereas a system with 5,000 detectors has a pixel size of 10 m. The effects of changing the pixel size of an image are shown in Figure 2.32. At 30 m resolution the approximately circular estate is very prominent and contrasts well with the more regular outline of surrounding fields. The pattern is maintained at 60 m (Figure 2.32b) though at 120 m much less detail can be discerned. Increasing the pixel size to 240 m effectively destroys any patterns that can be observed at the higher spatial resolutions.

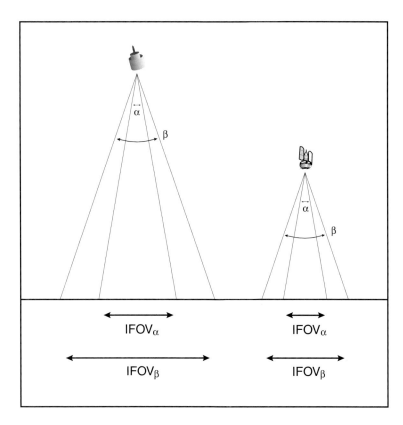

Figure 2.31 Effect of orbital height and viewing angle on the Instantaneous Field Of View. For a given height, the IFOV increases with viewing angle and for a given viewing angle the IFOV increases with height.

Landsat images (see Chapter 3) have been extensively used in remote sensing. The early images (1972–1982) had a resolution of about 80 m, though later images had a resolution of 30 m. Figure 2.33 shows the same area imaged at 80 m and 30 m and shows how much more detail can be determined with the better resolution system.

Choosing a Remote Sensing System

If one were presented with a choice between a remote sensing system (A) which had a spatial resolution of 10 m and obtained data for 10 bands at 1,024 grey levels and an 8-bit system (B) with a spatial resolution of 100 m for 4 bands, most people would assume that system A is the better of the two. How-

ever, before passing judgement on the systems it is important to ask yourself 'for what purpose will the images be used?' If the purpose is to investigate the location of major forest fires in Indonesia, then system B will be more than adequate for analysing such macroscale features. A number of other important points must also be borne in mind when one is determining which system best meets the particular needs of a researcher, such as the size of the data files and the signal-to-noise ratio of the system.

Size of Data Files

If one wished to image a 100 × 100 km area using one of these systems, then the size of the files holding the data is:

Modulation Transfer Function

An alternative measure of the fidelity of a lens involves the point spread function (PSF) and the modulation transfer function (MTF). In theory a uniform point source of light should produce a sharply defined edge where the intensity of illumination should drop to zero (Figure 2.30b). However, when imaged with a lens, the actual pattern observed is that of a gradual decreasing intensity rather than a clear boundary. This graph is known as the lens' point spread function. This blurring effect allows the modulation transfer factor M to be calculated. If a bar with a crenellate intensity function (Figure 2.30c) is photographed, it is possible to determine:

$$M_o \text{ (the object intensity modulation)} = \frac{I_{max} - I_{min}}{I_{max} + I_{min}}$$

$$\text{and } M_I \text{ (image intensity modulation)} = \frac{I_{max} - I_{min}}{I_{max} + I_{min}}$$

$$\text{The modulation transfer function } M = \frac{M_I}{M_o}$$

The MTF has a value between 0 and 1. If M is calculated for a range of bars with different spatial frequencies, a modulation transfer function curve is produced. A perfect lens has a regular decrease in image constant with spatial frequency due to increasing defraction effects. However, perfect lenses do not exist and in general lenses have the greatest fidelity at low spatial frequencies.

Although the above discussion has concentrated on the lens component of a camera system, the amount of detail that can be observed on a photographic image is also dependent on characteristics of the film. For example, a film consisting of large grains of silver halide will have a lower resolving power than a film with smaller grains. The resolving power of film ranges from about 25 lp/mm to 200 lp/mm. It is possible to produce an MTF for film and 'cascade' it with the MTF for the lens in order to produce an MTF for the imaging system.

System A:

$10^4 \times 10^4$ (number of pixels) $\times$ 4 (radiometric resolution in bytes) $\times$ 10 (number of bands) $= 4 \times 10^9$ bytes.

System B:

$10^3 \times 10^3$ (number of pixels) $\times$ 1 (radiometric resolution in bytes) $\times$ 4 (number of bands) $= 4 \times 10^6$ bytes.

The data file for system A is 1,000 times larger than for system B. Transmitting, storing, recording and digitally processing such a large file may not be cost-effective in terms of the extra information that is extracted compared with that obtained from the much smaller file of system B.

Signal-to-Noise (S/N) Ratio

The data that are recorded by a scanner's detectors are composed of the signal (which might be the reflectance from the ground) which is of concern to us and noise from aberrations in the electronics, moving mechanical parts or defects in the scanning system which degrade the signal to some extent. The noise component is a finite quantity which may be approximately constant for a particular system but the surface signal may vary. The surface signal obtained from a broad waveband (e.g. 0.4–1.0 μm) will be greater than that obtained for a narrow section of that waveband (0.4–0.5 μm); the signal-to-noise ratio is thus better for the broad waveband. As far as radiometric resolution is concerned, the steps between adjacent DNs must be greater than the DN produced by the

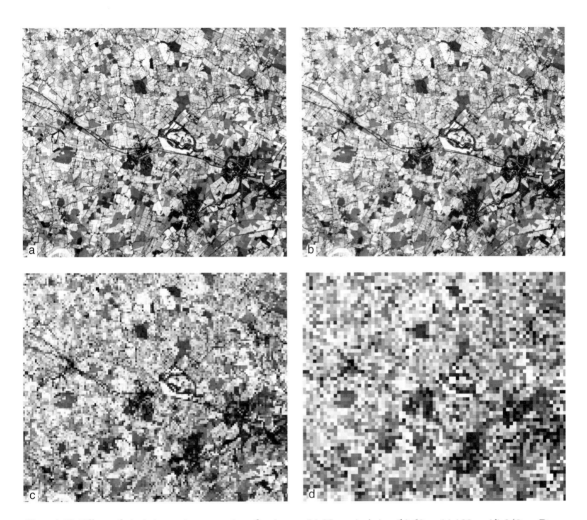

Figure 2.32 Effects of pixel size on interpretation of an image: (a) 30 m pixel size; (b) 60 m; (c) 120 m; (d) 240 m. Data courtesy of ERA-Maptec.

noise component otherwise the change in DN cannot confidently be attributed to spectral reflectance properties on the ground. The signal from a large pixel will also be stronger than that obtained from a small pixel. Thus increasing the spectral, spatial and radiometric resolutions of a system may decrease the signal-to-noise ratio to such a degree that the data may not be reliable. Some remote sensing systems allow the data to be obtained in either one of two modes: at a high spatial and low spectral resolution or at a high spectral and low spatial resolution. SPOT

can obtain one band of data with a 10 m spatial resolution or three bands of data with a 20 m spatial resolution. Another important aspect of the two systems regarding the signal-to-noise ratio is known as the dwell time, which is the time required for the detector IFOV to cross a pixel. In general, the longer the dwell time, the greater the signal-to-noise ratio. The dwell time for a pushbroom system is often much greater than for a transverse scanning system because the former has an individual detector allocated to each pixel. This enables a pushbroom system

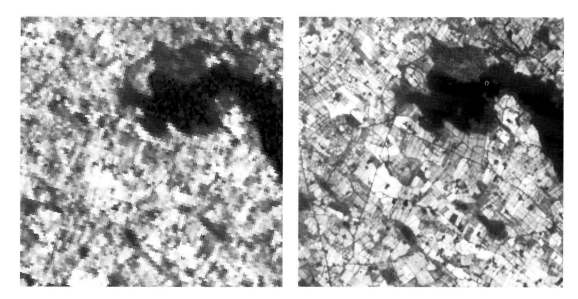

Figure 2.33 (a) 80 and (b) 30 m resolution Landsat images of the same region. Considerably more detail can be discerned on the 30 m resolution image.

to maintain a high signal-to-noise ratio with a high spatial resolution.

Detection and Resolution on Digital Images

A remote sensing system with a spatial resolution of 80 m will not resolve features smaller than this, but it may detect smaller features if the contrast between the object and the background is sufficiently great. Consider the situation shown in Figure 2.34, which illustrates a linear structure such as a road. If one were to measure the reflectances for all the features in different parts of the scene, the low digital numbers associated with the road (a DN of 7) would show that it is significantly darker than its surroundings. However, if a satellite images the same area, only one digital number can be assigned per pixel. The satellite recording system will assign the same DNs for the surrounding pixels as before but the DN for the pixels crossed by the road will be a function of the combined reflectances of all the features within that pixel. Thus a black road on a pale background will result in that pixel being assigned a DN which is higher than

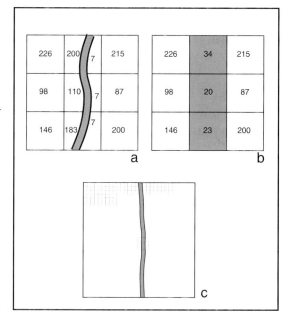

Figure 2.34 The difference between detection and resolution on a digital image. Narrow linear features with a high contrast to the background may be detected because they affect the digital numbers of the pixels they cross but might not be resolved.

the road but lower than the background and consequently that pixel may appear dark grey on the image. When the entire image is viewed, because the pixels are so small, the line of very dark pixels gives the impression that the road is being imaged whereas in reality it is a line of pixels that is being observed. Thus the presence of the feature is confirmed but the feature itself does not appear. Figure 2.35a shows a satellite image with a nominal resolution of 80 m. The presence of an east–west-trending road approximately 40 m wide can be detected because in this instance the road has a much brighter signature than the surroundings. Dark igneous dykes, which are narrower than the resolution, can be detected on a satellite image of the Sudan because they contrast well with the pale sandy background (Figure 2.35b).

2.5 SCALE OF AN IMAGE

One of the most important concepts regarding a remotely sensed image is that of scale. An image is a representation of a particular part of the Earth. In order to interpret the image properly it is important to know how relationships between different elements on the image vary compared to how the elements vary in reality. The scale of an image relates distances on the ground measured in any given units to distances on an image measured in the same units. Scale is commonly expressed as a ratio or as a fraction. Thus a scale of 1: 50,000, which is equivalent to 1/50,000, means that, if two features are separated by 1 mm on an image, then the two features are in reality 50,000 mm (50 m) apart. The scale of an image is given as a fraction by the formula:

$$\text{Scale} = \frac{\text{distance separating two features on an image}}{\text{distance separating the same two features on the ground}}$$

where the distances are given in the same units. (Note the actual unit does not affect the calculation, so that inches, centimetres or millimetres may be used.) However, it is important to realise that if image A and image B are the same physical size then the area covered by image B will be four times that of image A.

The scale of an image will change if the physical size of the image is altered. During the preparation of this book many images were examined at one scale but in order to be reproduced in the book, their size had to be altered and this changed their scale.

The scale at which imagery is commonly interpreted varies over a number of orders of magnitude.

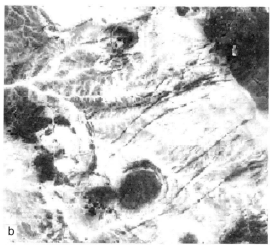

Figure 2.35 (a) Detection of an east–west-trending road (pale signature) which is narrower than the spatial resolution of this MSS image. (b) Detection of narrow dykes in the Bayuda Desert, Sudan. Image 35 km wide. Both the road and the dykes have a high contrast with the background.

Example: Two buildings which are 360 m apart are separated by 36 mm on image A and 18 mm on image B. Calculate the scale of image A and image B and state which of the two images has the larger scale.

For image A:

$$\text{Scale} = \frac{36}{360,000} = \frac{1}{10,000}$$

Therefore the scale of image A is 1: 10,000. Performing the same calculation for image B produces a scale of 1: 20,000.

Intuitively, one might expect a scale of 1: 20,000 to be greater than a scale of 1: 10,000, but, the opposite is the case. For any two images, the one whose distance between two features is closest to the distance between the features in real life will be at the larger scale. Thus in the example above, 36 mm (image A) is closer to the real distance, 360 m, than 18 mm (image B). Therefore image A is at a larger scale (1: 10,000) than image B (1: 20,000). Although image A is at twice the scale of image B, the area of the same feature on image A will be four times that seen on image B.

Example: Calculate the area in mm^2 that a rectangular building which is 80 m long and 50 m wide will be for image A and image B.

For image A with a scale of 1: 10,000, an 80 m length (80,000 mm) is equivalent to 80,000/10,000 mm or 8 mm. Similarly, 50 m is equal to 5 mm on image A. Thus the area of the building is 40 mm^2. A similar calculation for image B gives an area 10 mm^2 of (4 × 2.5). The area of the building is four times greater for image A than for image B.

Example: If image A and image B are both 20 × 20 cm what area is represented on the images?

For image A, with a scale of 1: 10,000, 1 cm is equivalent to 0.1 km in reality. Therefore the area covered by image A is 4 km^2 (2 × 2). The same 20 × 20 cm is equivalent to an area of 16 km^2 (4 × 4) for image B, which has a scale of 1: 20,000.

scales, 1: 5,000. National mapping programmes using aerial photography vary depending on the country. The United States Geological Survey mapped at a scale of 1: 40,000, each photograph covering an area of 9.2 × 9.2 km, whereas the Irish Geological Survey

Example: An image at a scale of 1: 15,000 measuring 24 × 24 cm is reproduced at a size of 18 × 18 cm. What is the scale of the reproduction?

24 cm on an image with a scale of 1: 15,000 is equivalent to 24 × 15,000 cm on the ground. If this length is then displayed as 18 cm on an image, then the image has a scale of:

$$\frac{18}{24 \times 15,000}$$

or 1: 20,000

photographed at a flying height of 4,500 m with a scale of 1: 30,000. In the latter case, each photographic negative is 23 × 23 cm and covers an area of approximately 6.9 × 6.9 km.

An assumption has been implicit in all the above discussions. It has been assumed that the scale for any particular image is constant and does not vary for different parts of the image or if measurements have been made in different directions. It will be shown later that this assumption is not valid for some remote sensing

Scales of 1: 1 million or 1: 10 million are common for satellite systems. Aerial photography is obtained over a range of scales. Thermal surveys are often obtained at low flying heights and at relatively high

systems. Scale distortions introduce inaccuracies in the image and are discussed in Chapter 3.

2.6 IMAGE INTERPRETATION PRINCIPLES

Using a remote sensing image to help in the understanding of a particular area may seem daunting at first because in some sense every image is unique. Images can encompass areas experiencing different climates, vegetation types, rock types and so on. Even images of the same area may change considerably depending on the time of year. In image analysis one can consider two levels of interpretation:

1 differentiation
2 identification.

Differentiation

Differentiation implies that features that can be observed on an image can be distinguished from each other without what the features represent being positively known. Although every image is different, all images share common characteristics, which often allow differentiation.

Tone

Panchromatic images exhibit tonal variations, which are related to the properties of the surfaces which are being imaged. Thus a camera will record different reflectances whereas a thermal sensor will distinguish areas with different thermal properties. It is possible to quantify the tone of a particular surface by making measurements of the amount of light that passes through the negative. However, interpreters generally employ a qualitative approach to image analysis and refer to a surface as being light, dark or even medium toned, which is essentially a comparative method depending on the impression of the range of tones that can be observed on the image. Thus a surface which may be described as dark toned if it is surrounded by paler surfaces may on another image be described as medium toned if darker and paler sur-

faces are in the region. Comparing tones between images may also be difficult because of differences in processing. Tonal relationships (assuming the same waveband is being sensed) do not change with scale variations. Thus a surface which is dark on a low-altitude aerial photograph will also appear dark on a high-altitude aerial photograph of the same area.

Texture

It is difficult to provide a rigorous definition of texture. Typical dictionary definitions refer to 'the arrangement of small constituent parts' or the 'representation of the structure and detail of objects'. In remote sensing terms, texture can be thought of as a subjective impression of the rate of change of tone for a surface. An object which maintains a constant tone throughout may be described as having an even texture. If the surface tone changes gradually it has a smooth texture, whereas a surface where the tone changes by large amounts over short distances may be described as having a rough texture. Texture, unlike tone, varies depending on the scale at which it is viewed. Thus a forest canopy consisting of different species at different heights at different stages in their growth cycle may have a rough texture on an aerial photograph obtained at one metre resolution by a low-flying aircraft but the texture for the same forest may appear smooth on an image obtained at 80 m resolution by a satellite.

Spatial Relationships and Context

Although tone and texture are important components in the interpretation of a remotely sensed image, the interpreter also brings a level of knowledge based on everyday living and a degree of experience which are often important. For example, consider the following description of a feature that is observed on an aerial photograph: 'a linear feature of constant width connecting two towns along which cars can be seen at intervals'. Most people will recognise this as a description of a road, although, neither the tone nor the texture of the feature was mentioned in the description. The road could have been formed of pale concrete or dark tar and could have had either a smooth or rough texture. However, the shape of the feature, its spatial relationship to other features in the image (i.e. the towns) and its context within the overall scene are all-important characteristics which aid in the interpretation of an image.

Identification

The combined tonal, textural, contextual and shape characteristics that constitute any feature observed on a remote sensing image are referred to as that feature's signature. If all features had a unique signature they could be unambiguously identified. However, this is not the situation in reality. The use of colour can greatly increase the potential for identification because two features with identical tones on a black and white image may have quite different colours. Although it may not always be possible to identify a feature with 100 per cent certainty, it is often possible by an examination of signatures to discount particular features with varying degrees of certainty. The differences between natural and artificial features in the landscape are often readily identified by remote sensing means. Cultural structures such as modern cities are often associated with straight lines and regular patterns of buildings and routeways. Airport runways are often seen as linear stripes or as a characteristic X pattern on remotely sensed images (Figure 2.36).

Many natural features have characteristic (often asymmetric) shapes which may help in any identification procedure. Glacial activity produces erosional features, mainly (but not exclusively) observed in mountainous areas. Amphitheatre-shaped hollows on the sides of mountains represent corries, the sites of former glaciers; where two formed on one mountain, erosion can produce an arete, which is a narrow ridge separating two corries. Depositional features such as eskers, moraines, outwash plains and drumlins may be formed by glacial activity. Apart from outwash plains, which from the air are often quite featureless, the other structures are elongated and may be observed on satellite images (see Figure 2.18b and accompanying WWW site).

Linear, topographic, tonal or textural features termed 'lineaments' may be observed on remotely sensed images. Lineaments are of particular interest to geologists because they often represent faults or fractures along which ore bodies may be concentrated. Rocks which are layered can often be picked out by remote sensing techniques because of distinctive colour or textural differences or because of differential erosion. Particular landforms such as volcanoes or dykes have characteristic shapes and indicate the presence of volcanic rocks. Many volcanoes have a classic conical shape and a central crater, which other features do not possess. Lava flows sometimes have a characteristic lobate form. Dykes are generally vertical layers of rocks and can appear as 'stripes' on an image (see Figure 2.35b). Two books (Drury 1998; Lillesand and Kiefer 1994) include numerous examples of the signatures of natural features obtained from airborne and spaceborne sensors and the reader is directed to these texts for a more comprehensive explanation and description of the appearance of these features on remotely sensing images.

2.7 HUMAN VISUAL SYSTEM

Although the characteristics of the sensors that obtain remote sensing data can influence the resultant image, information extraction from remotely sensed images rests overwhelmingly on visual interpretation. Indeed, many digital image processing techniques that are applied to remote sensing data are designed to accentuate differences on an image in order that they may be more easily examined visually. Consequently, knowledge of the workings of the human visual system is important for a full understanding of remotely sensed imagery.

The human eye is approximately spherical in shape with a diameter of 22 mm (Figure 2.37). It is essentially equivalent to a lens with a 17 mm focal length when it is viewing distant objects and 14 mm when observing nearby features. The human eye is only sensitive to electromagnetic radiation within the visible range of the electromagnetic spectrum (0.4–0.7 μm). Light enters the cornea and passes

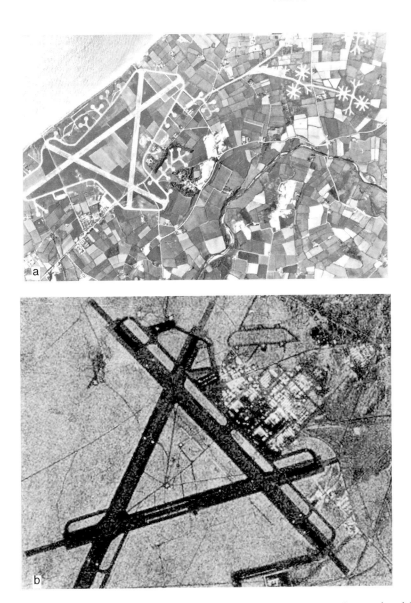

Figure 2.36 Typical X architecture of airport runways as observed on (a) an aerial photograph, reproduced from an Ordnance Survey aerial photograph with the permission of the Controller of Her Majesty's Stationary Office: Crown Copyright (permit number 966); (b) a Side Looking Airborne Radar image, Courtesy of Sandia Laboratories.

through the iris, which acts as an aperture regulating the amount of light that passes through the pupil and reaches the lens. For high-intensity radiation the iris reduces the pupil diameter to 2 mm and at low-intensity levels it opens to 8 mm. The amount of light that enters the eye can vary by a factor of 16 (i.e. 64/4), because the aperture is circular and its area varies as the radius squared. The light passes through the lens, which changes shape depending on the distance of the object from the

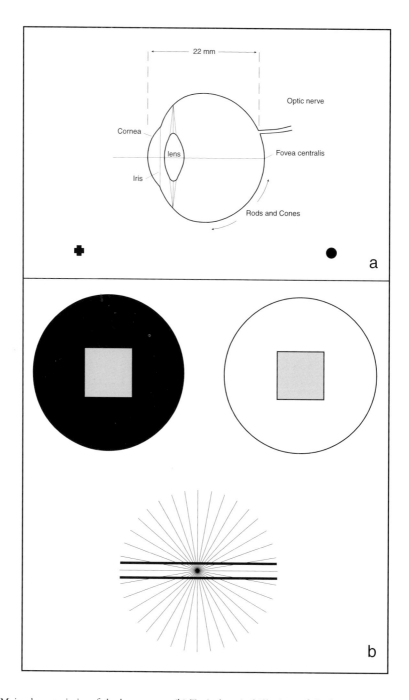

Figure 2.37 (a) Main characteristics of the human eye. (b) Typical optical illusions of the human visual system. The squares appear to be different shades of grey because of the different tonal backgrounds and the parallel lines appear to diverge in the centre of the wheel.

eye. The primary aim of this accommodation function is to focus the image onto the retina. The depth of range on which the human eye can focus ranges from about 10 cm to infinity though the former distance increases with age.

As mentioned above, the iris in the eye can change the intensity of light entering the eye by a factor of 16. If the eye were restricted to this factor we would have very poor visual capabilities. However, it can obtain information over a substantially greater range of light intensities (100,000) by the use of receptors on the retina. Two types of receptor, cones and rods, which react when stimulated by light, are located on the retinal surface. There are approximately 100 million long thin rods, with the highest concentrations near the periphery, and 10 million broad cones, which are most numerous around the fovea centralis. The receptors are analogous to the silver halide particles on a film. When they are exposed to light they react by means of a pigment alteration which sends a signal to the optic nerve via three types of nerve cells. The optic nerve transmits the signals to the brain, where they are interpreted and an image is formed. Thus one may have perfectly functioning eyes but brain damage can result in blindness. There are no receptors on the retina where the optic nerve is located, light falling on this portion of the retina will not be registered and as a result a blind spot exists. This can be confirmed by examining Figure 2.37. Hold the book at arm's length, close the left eye and look directly at the cross with your right eye. Slowly move the book towards you and at a certain distance the black spot will no longer be visible. If you move the book closer the spot will reappear. If the same experiment is performed with the right eye closed and the left eye open, the spot will not disappear. This is because the optic nerve exits on the opposite side of the retina for your left eye.

When a monochromatic photograph is being examined, the relative tonal differences, which are often related to different surface classes, can be determined. However, this distinction between areas of different tones is subjective and depends to some extent on the dominant tones in the vicinity. Figure 2.37b shows two square regions with similar grey levels surrounded by circular areas with markedly different grey levels. The square surrounded by the white circle appears darker to the eye than the square encircled by black. A similar effect can occur in colour images. If one region is dominated by a single colour then the complement of that colour will appear accentuated. Thus, in a blue region, yellow will appear more pronounced than red. Similarly, the horizontal lines on Figure 2.37b are parallel, though to our eyes they diverge in the centre of the diagram. This subjectivity does not apply to computer algorithms, which can statistically manipulate the digital data in order to produce classes with similar DN ranges.

The cones and the rods on the retina are sensitive to different stimuli. Cones react to colour differences whereas rods are most active in low light conditions. The concentration of cones at the fovea centralis means that visual acuity is optimised at this location. This allows us, in normal lighting conditions, to focus on fine detail. The lack of rods at the fovea centralis means that, if you want to examine something in conditions of very low illumination, you should not look directly at it but have it aligned about 15 degrees off centre. For example, after you have been outdoors at night for over an hour, the number of stars that you can observe are at a maximum because the pupil is at its maximum aperture and the rods at the retina's periphery are at maximum sensitivity. Very faint stars that are barely visible will disappear from view if they are stared at directly. Rods do not register colours, and therefore in very low lighting conditions, for example on a moonlit night, the human eye can only see in black, white and shades of grey. There are three types of cones which are sensitive to blue, green and red/orange light (Figure 2.38). The human eye does not respond uniformly over the visible light range. The green cones peak at 0.53, the red/orange at 0.58 and the blue at 0.44. The peak is highest for the green, slightly less for the red/orange and considerably lower for the blue. Thus the human eye, in good illumination conditions (photopic vision), is most sensitive to green variations, probably because we evolved in a world dominated by a Sun whose maximum emittance occurs in the green range. However, at low levels of illumination (scotopic vision), this maximum peak shifts to shorter wavelengths (Purkinje effect) and sensitivity to red light is reduced (Figure 2.38). The human eye can distinguish hundreds of thousands of colours. Colours in the vicinity of the green peak that are separated by 10^{-3} μm can be distinguished whereas in the blue and red ranges, a difference of 6×10^{-3} μm can be determined. Absolute measurement of radiance as measured by sensors, is referred to as radiometry whereas photometry is restricted to the visible range and takes account of the characteristics of the eye. Colour blindness, where a colour is perceived differently, can be explained by assuming the absence of cones which are sensitive to a particular colour. Although the human visual system may determine relatively subtle colour differences, it is a poor discriminator of grey levels. The maximum number of grey levels that can be seen is about 30. Figure 2.39 illustrates the same area shown in 2, 4, 8 and 32 grey levels. The difference between 2 and 4 grey levels may be determined but it is difficult to discern major differences between 8 and 32 grey-level images.

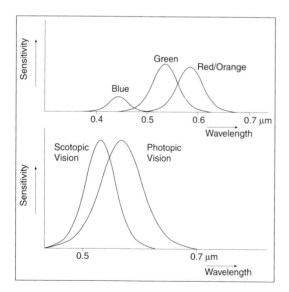

Figure 2.38 Sensitivity of the human visual system to blue, green, and red/orange light and at different levels of illumination.

Binocular Vision

When viewing an object, we intuitively expect that it will appear smaller at greater distances. In effect, the angle that the light rays make as they converge in the eye is smaller for larger distances. Humans, like the majority of mammals, have two eyes facing forward. Thus both eyes view the same object simultaneously, but because the eyes are separated by a short distance (6.5 cm), each eye will produce a slightly different record of the object. The separate images recorded by the eyes are fused within the brain in order to form a single image which will have a three-dimensional appearance. Such binocular vision is best achieved when both eyes are of approximately equal strength. Some people have difficulty experiencing binocular vision because one eye is much stronger that the other. In this situation, in order to avoid double vision, the brain suppresses one image. To ascertain whether you have binocular vision hold a pen in a vertical position at arm's length and line it up (while keeping both eyes open) with a vertical feature such as the edge of a window or building. Then alternately close each eye and the pen should be seen to move to either side of the vertical feature. If you close your right eye and the pen and the pole remain aligned, then your left eye is dominant. As the distance of a feature from an observer increases, the angle between the light waves entering the eyes decreases. The minimum angle that can be subtended is about 30 seconds of an arc; the eyes cannot therefore produce three-dimensional images beyond about 300 m. The lack of depth perception at this distance means than an observer on an aircraft would not be able to appreciate small variations in ground topography. This disadvantage can be overcome by using artificial aids in the analysis of aerial photographs (see Chapter 3).

2.8 CHAPTER SUMMARY

- All electromagnetic radiation moves at a constant speed in a vacuum but the wavelength of the

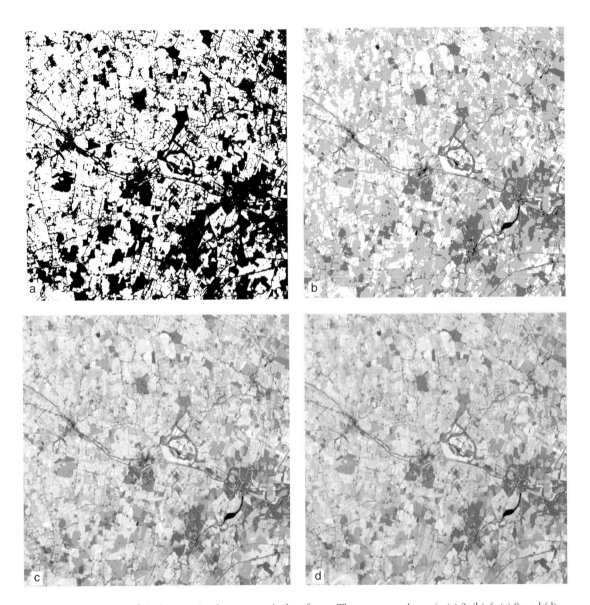

Figure 2.39 Sensitivity of the human visual system to shades of grey. The same area shown in (a) 2; (b) 4; (c) 8; and (d) 32 grey levels. Data courtesy of ERA-Maptec

waves varies. Visible light occupies a very small part of the electromagnetic spectrum.

- A passive remote sensing system uses the Sun as its source of illumination whereas an active system has an onboard source of electromagnetic radiation.

- The atmosphere is composed of gases and aerosols which selectively absorb and scatter electromagnetic radiation at particular wavelengths.

- A number of atmospheric windows exist at particular wavebands through which radiation may pass. The most important atmospheric windows are in the visible/near infrared and microwave ranges and in parts of the thermal infrared (3–5 μm and 8–14 μm).

- Scattering by particles in the atmosphere can contribute to the signal measured by remote sensing sensors. Selective scattering (such as Rayleigh or Mie scattering) is wavelength dependent and depends also on the relative size of the particles compared with the wavelength of the radiation. Non-selective scattering is not dependent upon wavelength.

- Vegetation has a low reflectance (and transmission) in the visible range which increases greatly in the infrared. Changes in the signature of vegetation may be more apparent in the infrared before any changes within the visible range can be observed because of the greater response of green vegetation to the near infrared radiation.

- Water has a low reflectance in the visible range and very low in the infrared range of the spectrum. Ice has a high reflectance in the visible. Transmission through water is greater for shorter wavelengths. Blue light can thus penetrate farther than red light.

- There are a number of different types of remotely sensed images, each with its own advantages and disadvantages. Panchromatic images are recorded in black and white format within the visible range of the electromagnetic spectrum. They are cheap and a substantial amount of back-coverage exists for most countries but it may be difficult to distinguish different surface classes. Black and white infrared is less affected by haze than panchromatic photographs and is particularly useful for vegetation surveys or water-body detection. Natural colour images have a lower definition than black and white images but the observed colours accord with our own visual system, which makes interpretation easier. Multispectral images are usually acquired separately in the blue, green, red and infrared. Such systems are very versatile as images can be examined individually or as composites.

- A false colour image can be produced by projecting the green image in blue, the red image in green and the infrared image in red. Spectral differences are often more apparent on a false colour composite.

- A digital image is a regular grid array of squares (or rectangles) where each square is assigned a digital number (DN) which is related to some parameter (such as reflectance or temperature) which is being measured by a remote sensing system's sensor.

- Passive digital remote sensing systems obtain their data mainly by means of a transverse scanner or by using a pushbroom system. A transverse scanning system is an electro-mechanical device that obtains data from narrow swaths of terrain by means of a scanning mirror which sweeps across the scene and then directs this reflected (or emitted) radiation towards the onboard detectors. A pushbroom system does not employ a mirror but has a separate detector for each pixel in the line.

- Four types of resolution can be recognised in remote sensing. Spectral resolution is a measure of the number and width of bands sensed and temporal resolution is a measure of how often imagery of a particular area is obtained. In digital images, the number of grey levels that are measured is the radiometric resolution while the spatial resolution is essentially a measure of the smallest features that can be observed on an image.

- Images are interpreted by eye and the human visual system does not have a uniform response at all visible wavelengths. The eye is most sensitive to green light and least sensitive to blue light.

- The scale of an image relates distances on the ground measured in any given units to distances measured on the image in the same units of measurement.

- It may not always be possible to make a unique identification of the features observed on an image, but variations in tone and texture and the spatial relationships of the observed features can be used as an aid in interpretation.

SELF-ASSESSMENT TEST

1 What is the wavelength of electromagnetic radiation which has a frequency of 5×10^{10} Hz? What type of electromagnetic radiation has this wavelength?

   $c = 3 \times 10^8$ m/s.

2 Why is clear non-turbulent water blue/green in the visible part of the spectrum and black in the near infrared?

3 What are the two forms of selective scattering and how does selective scattering differ from non-selective scattering?

4 What is a false colour image and how is it produced?

5 Which components of the atmosphere produce absorptions at (1) 1.4 μm; (2) 2.7 μm and (3) 6.3 μm?

6 Two buildings are 54 mm apart on a 1: 30,000 map and 85 mm apart on an aerial photograph. What is the scale of the photograph? What is the true length of a road which measures 15 mm on the aerial photograph?

7 What is the difference between an active and a passive remote sensing system? In which part of the electromagnetic spectrum do active remote sensing systems usually operate? Why is this?

8 What is the most significant feature of the spectral signature of vegetation in the 0.4–1.1 μm waveband?

FURTHER READING

Avery, T. E. and Berlin, G. L. (1992) *Fundamentals of Remote Sensing and Airphoto Interpretation*, 5th edition, Englewoods Cliffs, N J: Prentice-Hall. (Chapter 1)

Campbell, J. B. (1996) *Introduction to Remote Sensing,* 2nd edition, London: Taylor and Francis. (Chapters 2 and 3)

Drury, S. A. (1993) *Image Interpretation in Geology*, London: Chapman and Hall. (Chapters 1, 2 and 4)

Lillesand, T. M. and Kiefer, R. W. (1994) *Remote Sensing and Image Interpretation*, New York: John Wiley and Sons. (Chapter 1)

M^{C}Cloy, K. R. (1995) *Resource Management Information Systems: Process and Practice*, London: Taylor and Francis. (Chapter 2)

3

REMOTE SENSING SYSTEMS

Chapter Outline

3.1 Images obtained from aerial systems
Types of aerial photograph
Scale of vertical aerial photograph
Distortions on an aerial photograph
Stereoscopic viewing
Thermal imaging
Airborne hyperspectral imaging
Multisensor aerial surveys

3.2 Images obtained from low-orbiting satellite platforms
Landsat system
SPOT system
NOAA meteorological satellites
DMSP
JERS-1
Imaging in the microwave
Radar: theoretical concepts and systems
Other low-orbiting spaceborne remote sensing systems

3.3 Images obtained from geostationary satellites

3.4 Other forms of remote sensing

3.5 Remote sensing – developments in the future

3.6 Chapter summary

Self-Assessment Test

Further Reading

3.1 IMAGES OBTAINED FROM AERIAL SYSTEMS

Types of Aerial Photograph

Aerial photographs can be obtained by a camera system that points vertically down at the landscape (vertical photographs) or one that points to the side or in front of the aircraft (oblique photographs) (Figure 3.1a). A low oblique photograph images the terrain to the side or in front of the aircraft whereas a high oblique one extends to the horizon. National mapping programmes invariably obtain vertical photographs because they can be examined stereoscopically and scale distortions can be corrected relatively easily. A vertical aerial photograph is typically 23 × 23 cm square (9 × 9 inches) and has a number of characteristics (Figure 3.1b). Four tick marks termed fiducials are located at the four corners (or at the centre of the four sides) of the photograph. Lines joining opposite fiducials intersect at the centre of the photograph, which is termed the principal point (P, Figure 3.1b). The camera lens is immediately above this point when the photograph is taken. Various types of information are often located along the edge of the photograph such as the time the photograph was obtained, the altitude and attitude of the aeroplane and a serial number (Figure 3.1b). A vertical aerial photograph is not a map. Spatial relationships that are constant on a map are

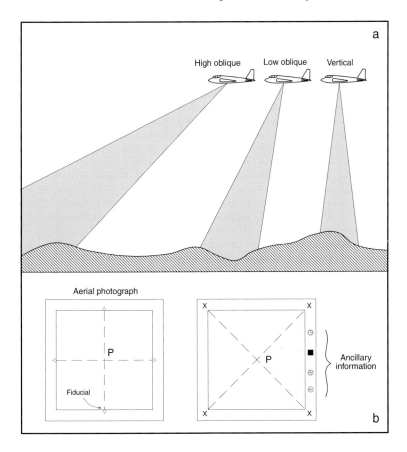

Figure 3.1 (a) Types of aerial photograph. Vertical photographs are the most common. (b) General layout of vertical photographs, (P is the principal point).

not constant on an aerial photograph, as will be discussed below.

Scale of a Vertical Aerial Photograph

The scale of an image was defined in Chapter 2 as the distance separating two features on the image divided by the distance separating the same two features on the ground. It is possible to evaluate the scale of an aerial photograph by using the focal length of the camera system (f) and the height (H) above the terrain at which the photograph is obtained. It is given by the expression:

$$\text{Scale} = \frac{\text{focal length of camera}}{\text{flying height above terrain}} = \frac{f}{H}$$

equation 3.1

Example: Calculate the scale of a vertical aerial photograph which was obtained for a level area 250 m above sea level by an aircraft flying at a height of 1,500 m above sea level. The camera system has a focal length of 10 cm.

H, the height above the terrain, $1,500 - 250$ is 1,250 m and f, the focal length, is 0.1 m. From equation 3.1:

$$\text{Scale} = \frac{0.1}{1,250}$$

$$= 1/12,500$$

Therefore the scale of this photograph is 1: 12,500

Distortions on an Aerial Photograph

In the discussions of the concept of scale s o far, it has been assumed that the scale for any image obtained by a remote sensing system is constant. However, equation 3.1 shows that the scale of an image will vary depending upon the height at which the photograph is taken. Consider the situation shown in Figure 3.2, where an aircraft in level flight 2,000 m above sea level, with a camera focal length of 10 cm, is flying over terrain with a elevation of 0 m and 1,000 m. An aerial photograph of this scene will have a different scale at different elevations. For the upland region (1,000 m) the scale from equation 3.1 is 0.1/1,000 or 1: 10,000. At sea level the scale is 0.1/2,000 or 1: 20,000. Thus two features separated by 10 mm on the photograph in the upland area are in reality 100 m apart whereas two features at sea level, also 10 mm apart on the photograph, are 200 m apart. The scale given for an aerial photograph is often an average scale. The example given here is quite extreme. However, it illustrates the important point that a vertical aerial photograph cannot be considered analogous to a map, where the scale is constant everywhere.

A second distortion that is evident on vertical aerial photographs is that of the radial distortion of tall structures. A tall feature, such as a chimney, if it is at the principal point (1, Figure 3.3a) when the photograph is taken, will appear as a single dot on the corresponding photograph (1, Figure 3.3b). The base and top of the feature will be coincident and no impression of height will be obtained. If, however, the feature is not at the principal point when the photograph is obtained (2, Figure 3.3a), the top and the base of the feature will not be coincident and it will appear to lean away from the principal point on an aerial photograph. This radial displacement increases as distance from the principal point increases and is most noticeable around the edges of a vertical photograph. This effect may make the construction of mosaics difficult, as the same feature will be leaning in different directions in adjacent aerial photographs. However, an advantage of this radial displacement is that the height of tall features may be evaluated by the equation:

$$h = \frac{d\,H}{r}$$

equation 3.2

where h is the true height of the feature, H is the flying height above the surface, d is the length of the feature as measured on the photograph and r is the distance from the principal point to the top of the feature. If d and r are measured in the same units – such

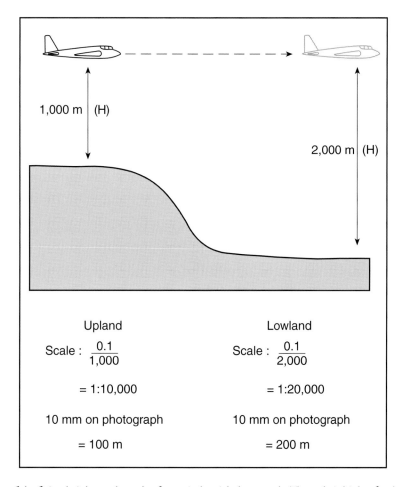

1,000 m │ (H)

2,000 m │ (H)

Upland

Scale : $\dfrac{0.1}{1,000}$

= 1:10,000

10 mm on photograph

= 100 m

Lowland

Scale : $\dfrac{0.1}{2,000}$

= 1:20,000

10 mm on photograph

= 200 m

Figure 3.2 Effect of the flying height on the scale of a vertical aerial photograph. The scale is higher for the higher ground.

as millimetres – and H is in metres, then the height of the object will also be given in metres. The parameters d and r also need to be measured parallel to each other.

Other distortions may also affect aerial photographs. The relationship between the negative and the terrain on Figure 3.3a show that they are parallel to one another. However, the aircraft is not a perfectly stable platform and may roll, pitch and yaw. Thus the camera system may not be parallel to the ground when the photograph is taken; this introduces tilt distortions. Variations in flying height caused by vertical movements of the aeroplane may affect comparisons

Example: Calculate the height of a tower which is observed to be 5 mm long and 200 mm from the principal point on a vertical aerial photograph. The flying height of the camera system is 2,000 m.

From equation 3.2:

$$h = \frac{5 \times 2,000}{200}$$

$$= 50 \text{ m}$$

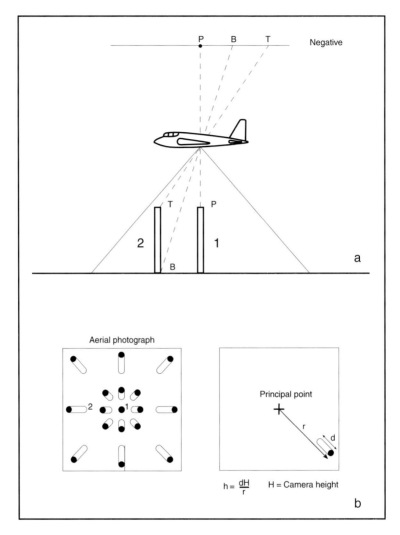

Figure 3.3 Radial distortion on aerial vertical photographs. Tall features are seen to lean away from the principal point. This distortion is greatest near the edge of the photographs.

between separate photographs. The Earth's surface is curved not planar and while this effect is negligible for small areas, on small-scale images obtained at high altitudes the effect is more pronounced.

Stereoscopic Viewing

A single vertical aerial photograph is a two-dimensional representation of a three-dimensional landscape. Because of their binocular visual capabilities, humans are able to assimilate three-dimensional information. Although slopes and shadows on an aerial photograph provide indirect clues to topographic variations, it is possible to use artificial aids to help in the interpretation. The variations in distance (or height, from the vertical viewpoint of an aircraft) that can be determined by the human visual system, assuming the human eye can detect objects which

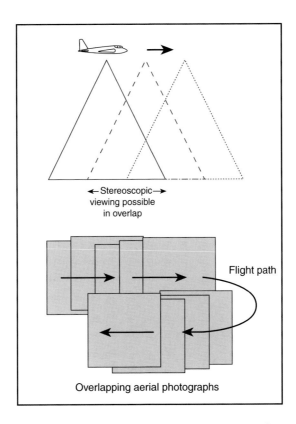

←Stereoscopic→
viewing possible
in overlap

Flight path

Overlapping aerial photographs

Figure 3.4 Production of stereoscopic aerial photographs by taking photographs with a 60 per cent forward overlap. Adjacent flight paths overlap by about 20 per cent.

subtend a 30 second angle of an arc, are given by the equation:

$$h = \frac{H^2 + 0.25\ b^2}{H + 6.67b \times 10^3}$$ equation 3.3

where h is the height, H is the viewing distance and b is the distance between the eyes (eye-base). Kilford (1979) provides a full derivation of equation 3.3. At a distance of 100 m, the human visual system will be able to discern a change of distance of 18 m. However, from an aircraft flying at a height of 2,500 m, only topographic variations greater than 2,100 m will be distinguishable by the human eye. However, if the eye-base (b) could be increased to 500 m instead of 65 mm, then height differences of 2 m could be determined at a height of 2,500 m. This effect can be achieved by regulating the speed of the aircraft and the rate at which the photographs are obtained in order to produce an overlap of 60 per cent in the direction of motion and a 30 per cent sidelap between adjacent strips (Figure 3.4). Two adjacent photographs can then be viewed through a stereoscope such that the left eye observes one photograph and the right eye sees only the other photograph (Figure 3.5). Stereoscopes vary in complexity: the simplest one, a pocket lens stereoscope, consists of plastic lenses with a magnification of about 2 placed on 12 cm-long retractable legs. The instrument is light and collapsible, and is thus ideal for taking out into the field, though a drawback is that only a small area can be viewed. This problem can be overcome by using a mirror stereoscope. Light rays reflected from

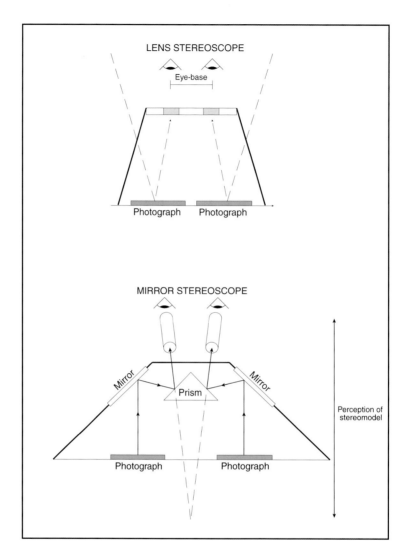

Figure 3.5 The use of a lens and mirror stereoscope to produce a three-dimensional effect for the overlap between adjacent vertical aerial photographs.

the photographic prints do not take a direct route to the eyes as in a lens stereoscope, but reach the eyes after being reflected from angled mirrors and a prism (Figure 3.5). This approach widens the viewer's eye-base by a factor of about 3 and consequently a much larger area may be examined stereoscopically. The brain is able to fuse the two separate images seen by each eye in order to produce a single three-dimensional scene. The eye-base for the stereoscopically observed scene is equivalent to the left and right eyes being positioned where the film was exposed, i.e. at the principal point of the photographs. This eye-base distance can vary depending on the speed of the aircraft but is typically of the order of 500–1,000 m and from equation 3.3 even quite small height differences may be discerned.

Vertical Exaggeration

A common characteristic of photographs being viewed stereoscopically is the vertical exaggeration that is observed. Shallow gradients appear steep and steep gradients seem extremely steep. This has the advantage that, even in areas of relatively low relief, topographic variations may be observed. The approximate amount of vertical exaggeration (v) is given by the formula:

$$v = \frac{d}{H} \times \frac{L}{b} \qquad \text{equation 3.4}$$

where d is the distance between the principal points of two adjacent photographs; H is the flying height above the terrain, b is the eye-base (6.5 cm) and L is the distance to the perception of the stereomodel. L/b is approximately constant; thus the vertical exaggeration is directly proportional to d/H, the base–height ratio.

A stereopair, which may be examined by means of a lens stereoscope, is illustrated (Figure 3.6). The most prominent feature on Figure 3.6 is the Fair Head Sill, a geological igneous intrusion that forms the high relief area. The sill terminates seaward as a steep cliff, 80 m above the scree, which forms low ground, adjacent to the shore. Note the presence of parallel fractures on the sill and how the illumination direction can be determined by the prominent shadows near the coast. (Applications of aerial photography are discussed in Chapter 4.)

Orthophotographs

An aerial photograph contains radial and terrain distortions. However, it is possible to correct these distortions, either by using a device called an orthophotoscope or by digital means. Such a corrected aerial photograph is termed an orthophotograph. In order to produce an orthophotograph digitally, a number of procedures must be followed. The aerial photograph

Figure 3.6 Stereoscopic photograph of the Fair Head Sill which can be viewed with a lens stereoscope. Reproduced from an Ordnance Survey aerial photograph with the permission of the Controller of Her Majesty's Stationery Office. Crown Copyright (permit number 966)

must be in a digital format. This can be achieved either by obtaining an image with a scanner system or by scanning the photograph. Next, the co-ordinates of easily identifiable features that can be seen on the image (which is distorted) are collected and the correct co-ordinates of the same features are also obtained from a map. Using these ground-control points, the computer produces a number of equations that transform the location of all the pixels on the distorted image to a properly orientated image. In order for this transformation process to be accurate, it is also necessary to input a file (called a digital elevation model) which stores the height of all the scanned pixels. If such a file does not exist, it is possible to create one by using a computer program to extract one for the area covered by two overlapping stereo aerial photographs.

Thermal Imaging

The basis of passive imaging systems is that sensors detect and measure the electromagnetic radiation reflected or emitted from different surfaces. For systems that operate in the visible and photographic infrared, the absorption characteristics for different bodies are wavelength specific. A body which absorbs a large amount of the electromagnetic radiation that is incident on it will correspondingly reflect little at that particular wavelength. However, two surfaces may have very similar reflectance characteristics within the visible and photographic infrared, but because they have dissimilar thermal properties they may be distinguished by sensors tuned to take measurements in the thermal infrared. A plethora of technical terms relevant to thermal imaging exist, many of which are interrelated, and the reader is directed to Sabins (1997) and Open Universiteit (1989) for a fuller discussion.

The thermal capacity (c) of a body is a measure of how much energy has to be added to it in order to raise a unit weight by one degree Celsius. Most metals have very low values of c (approximately 0.09) while many natural surfaces like vegetation, rocks and soils have a thermal capacity value of approximately 0.2. An exception is water, which has a c value of 1. The thermal conductivity (K_t) of a body is a measure of how quickly heat can move through it. Metals such as iron or aluminium have high K_t values (0.9). Natural substances such as water, rocks and soils have very low values of K_t (< 0.01). The combination of variations in thermal capacity (c), thermal conductivity (K_t) and density (ρ) for a body can be combined to give the thermal inertia (P), which is defined as:

$$P = \sqrt{(K_t \, c \, \rho)} \qquad \text{equation 3.5}$$

In a diurnal heating cycle, a body with a low thermal inertia will heat up to a high temperature during the day but will fall to a correspondingly low temperature at night, whereas a body with a high thermal inertia will experience a much more stable temperature regime: it will be relatively cool during the day but also relatively warm at night (Figure 3.7). Metals have high values of thermal inertia (0.9) because of their high density and high thermal conductivity, so that temperature variations between day and night are minimal. Consequently, on a daytime thermal image metals tend to be relatively cool (dark on the image) and may have a similar dark signature on a night-time thermal image. Soils and rocks have low thermal inertias (0.04) and will thus appear bright on a daytime image but dark if the image is obtained at night. Trees generally appear relatively cool (dark) compared with their surroundings during the day but warmer (brighter) at night, whereas grass generally changes from warm to cool going from day to night. Water has a thermal inertia similar to that of many rocks (0.036). However, its diurnal response is different from that of rocks because convection processes operate efficiently in water and not in rocks. Water generally appears very dark (cool) on a daytime thermal image but is brighter than its surroundings at night. The time of day at which the maximum temperature is achieved (t_1 to t_4 on Figure 3.7) varies and depends on the thermal properties of the body.

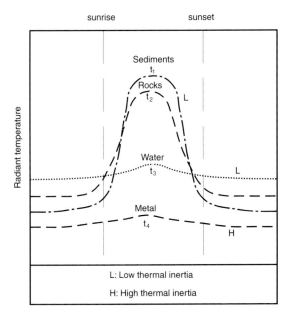

sunrise   sunset

Radiant temperature

Sediments
t_1

Rocks
t_2

L

Water
t_3

L

Metal
t_4

H

L: Low thermal inertia

H: High thermal inertia

Figure 3.7 Diurnal radiant temperature variations for low and high thermal inertia substances. 't' represents the maximum temperature. Note that the time at which the maximum temperature occurs varies for different substances.

From the foregoing discussion it is apparent that the time of day at which a thermal image is obtained will control the relative tonal signatures of the features within the scene. An image obtained at mid-day will be different from an image obtained at midnight. Daytime thermal images are strongly influenced by topographic variations because of the differential heating effects on slopes which face the Sun and slopes which are in shadow. Shadow effects caused by buildings are also prevalent in cities for daytime thermal imaging. Pre-dawn images tend to show features with different thermal properties rather than the effects of topographic shadows. It may not always be possible to acquire night-time thermal images. Airborne multispectral scanners which obtain data simultaneously in visible and thermal infrared wavelengths will need to operate during daylight if visible and infrared data are to be collected simultaneously. The data obtained on a thermal-imaging survey may also be degraded by transient weather conditions. Cloud shadows and wind may have a cooling effect on those parts of the scene which are affected by them and an image obtained after a rain shower may also appear darker.

Emissivity

The concept of a blackbody was briefly introduced in Chapter 2. A blackbody is one that absorbs all the incident energy that falls on it and obeys the Stefan–Boltzmann Law (see equation 2.2). Natural bodies do not obey this law and a constant, termed the emissivity (ε), has to be introduced into the equation, thus:

$$M = \sigma \varepsilon T^4 \qquad \text{equation 3.6}$$

where M is the energy emitted (referred to as the emittance or exitance), T is the temperature measured in degrees Kelvin, ε is the emissivity and σ is the Stefan–Boltzmann constant, which is equal to 5.6697×10^{-8} W m^{-2} K^{-4}.

The emissivity is defined as:

$$\frac{\text{energy actually emitted by unit area of a surface in unit time at a given temperature}}{\text{energy emitted by unit area of a blackbody in unit time at the same temperature}}$$

The emissivity of a blackbody is equal to 1 and all other materials have a value less than 1. Emissivity is wavelength dependent: a substance may have a particular ε value at one wavelength and a different value at another.

Example: From your everyday experiences, how do you think the emissivity of a rock such as granite compares with the emissivity of a metal such as aluminium?

The emissivity of granite is much greater than the emissivity of aluminium. Consider constructing a parabolic 'mirror' from aluminium and granite and placing a thermometer at the focus of both mirrors. The temperature measured by the thermometer at the focus of the aluminium mirror will be higher than for the rock because the aluminium reflects a large proportion of the incident energy. However, because the aluminium reflects a large proportion of the incident energy, it does not absorb a large amount. The granite reflects little and absorbs a larger amount of the incident energy. A blackbody that has an emissivity of 1 absorbs all the incident energy; the granite is thus more like the blackbody than the aluminium and consequently has an ε value closer to 1 than the ε value of the aluminium. The emissivity of granite is about 0.85 and the emissivity of aluminium is 0.05.

Example: An emittance of 450 W m^{-2} is measured for granite by a remote sensing radiometer. What is the temperature of the granite? From equation 3.6:

$$M = \sigma \varepsilon T^4$$

$$\text{therefore } T^4 = \frac{M}{\sigma \varepsilon} = \frac{450}{0.85 \times 5.6697 \times 10^{-8}}$$

$$= 93.4 \times 10^8$$

$$\text{therefore } T = \sqrt[4]{(93.4)} \times 10^2 = 310 \ ^\circ K$$

An aluminium surface which produced the same value of emittance would need to be at a temperature of $631 \ ^\circ K$.

So far the term 'temperature' has been used in this book without any qualification. If a temperature reading for the granite discussed above is obtained by placing a thermometer in direct contact with the rock at the same time as a remote sensing radiometer measures the temperature, the two results would be different. The radiometer measures what is known as the radiant temperature of the body whereas the thermometer in direct contact measures the kinetic temperature of the body. For a theoretical blackbody, the radiant and kinetic temperatures are equal, but for natural materials they are related thus:

$$T^4 \text{ (radiant)} = \varepsilon T^4 \text{(kinetic)} \qquad \text{equation 3.7}$$

Because ε is always less than 1 for a natural material, the radiant temperature is always less than the kinetic temperature. For materials with high values of ε such as water (0.99) the two temperatures are very similar. For the granite and aluminium discussed above, if the kinetic temperature for both was $400 \ ^\circ K$, then the radiant temperature of the granite would be $384 \ ^\circ K$ and that of the aluminium would be $189 \ ^\circ K$.

Acquisition of Thermal Images

Airborne thermal images are obtained by transverse scanning systems in which radiation from the ground is directed by means of a scanning mirror onto a detector which measures the emittance in a particular waveband. Although the thermal part of the electromagnetic spectrum extends from 3 μm to 1,000 μm, the detectors are designed to obtain measurements within atmospheric windows, most commonly 3–5 μm or 8–12 μm. In order to prevent spurious signals being recorded, the detector is shielded and maintained at a very low temperature (approximately 70 K). Variations in the thermal infrared signal are amplified and converted either to intensity variations for a glow tube which is used to expose photographic film or more commonly to an electrical signal which is recorded digitally. The differences on the resultant monochromatic image represent variations in the measured emittance. It is often assumed that grey variations on such a thermal image show the variations in temperature of the different surfaces in the scene. However, equation 3.6 shows that the emittance is also dependent upon the emissivity (ε): the tonal variations will be indicative of temperature

Plate 1.1 (a) A natural colour photograph is one of the simplest forms of remote sensing. Information about colour, texture and spatial relationships may be determined. (b) Alternative smaller scale version of 1.1a. Information from a larger area can now be obtained but detail about the figure is now reduced.

Plate 1.2 (a) Image of nearly an entire hemisphere which can provide information for a large region but shows little detail; acquired by the Meteosat geostationary satellite. (b) The British Isles obtained by a NOAA polar-orbiting satellite. Meteosat and NOAA images courtesy of National Remote Sensing Centre, UK

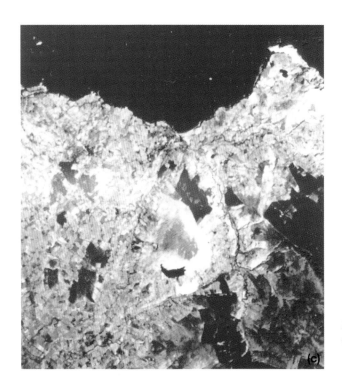

Plate 1.2 (c) Part of (b) shown with a greatly improved spatial resolution using the Landsat satellite.

Plate 1.2 (d) An even better spatial resolution is obtained on this aerial photograph but the area covered is much smaller than that shown in (c).

Plate 2.1 (a) Colour additive process, in which red and green combined produce yellow. (b) Colour subtractive process: a yellow filter passes red and green light but subtracts blue light.

Red

Yellow   Magenta

Green   Blue

Cyan   **(a)**

Cyan

Blue   Green

Magenta   Yellow

Red   **(b)**

Plate 2.2 Colour infrared aerial photograph displaying the characteristic red signature for healthy living vegetation.

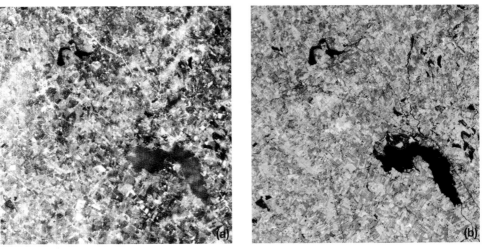

(a)   (b)

Plate 3.1 (a) An approximately true colour image produced using TM 1 projected in blue, TM 2 projected in green and TM 3 projected in red. (b) False colour image produced by projecting TM 3 in blue, TM 5 in green and TM 4 in red. See text for discussion. Approximately 15 × 15 km. Data courtesy of ERA-Maptec.

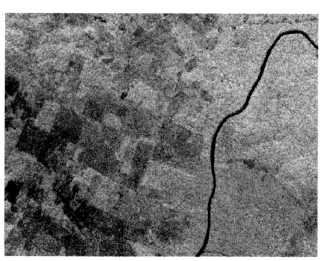

Plate 3.3 A false colour radar image of Brazilian forest formed by projecting an L band in HV mode in red, a C band in HV mode in green and an X band in VV mode in blue. Pink regions represent rainforest still in its natural state whereas regions where the forest has been felled and cleared for agricultural purposes are illustrated in shades of green and blue. The bright red is due to scattering from a transient rain shower. Courtesy of JPL/NASA.

Plate 3.2 SPOT multispectral image obtained in 1992 of part of the Caspian Sea coast near Gurjev. Width of image approximately 8 km. Copyright CNES 1992, courtesy of SPOT IMAGE.

Plate 3.4 An SIR-C image of the Karakax Valley in China draped over a digital elevation model produced using interferometry techniques. Courtesy of JPL/NASA.

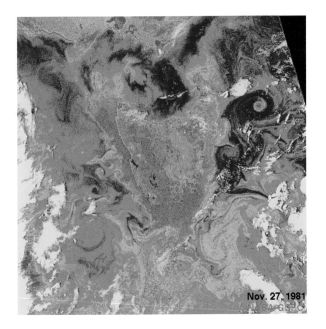

Plate 3.5 Coastal Zone Colour Scanner image showing the phytoplankton concentrations around Tasmania. The lowest concentrations are shown in blue/purple; medium concentrations are green/yellow and the highest are orange/red. Image courtesy of JPL/NASA.

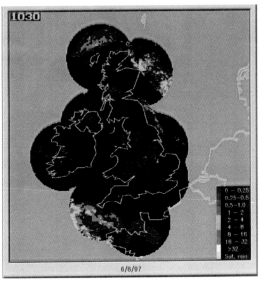

Plate 4.1 A composite of radar data and the overall rainfall analysis resulting from combination of satellite-derived rainfall and numerical weather prediction model data from the UK Meteorological Office Nimrod automatic forecasting system. British Crown copyright, reproduced with permission of the Controller of HMSO.

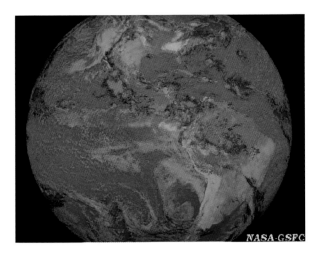

Plate 3.6 False colour image obtained by GOES-8. Such images allow global wind patterns and the movement of hurricanes and other major storms to be tracked. Images courtesy of Dennis Chesters GSFC/NASA.

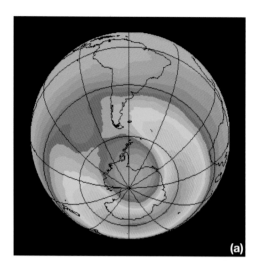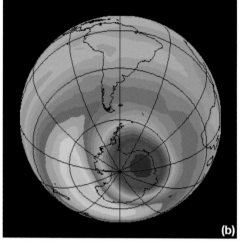

Plate 4.2 Ozone concentrations over Antarctica shown by TOMS/GOME in (a) 1979 and (b) 1992. Low values are shown in blue/purple and high values in red/orange. A marked decrease in ozone concentration between these two dates is apparent. Courtesy of JPL/NASA.

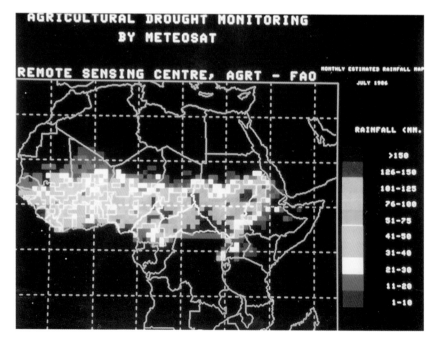

Plate 4.3 An ARTEMIS (African Real Time Environmental Monitoring Information System) product map. This is a system developed by the UN FAO (Food and Agriculture Organisation) which produces maps of rainfall estimates for 2.4 km squares covering the African continent routinely from Meteosat data. Courtesy of UN FAO Environment and Natural Resources Service.

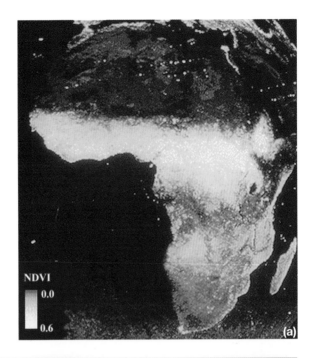

Plate 4.4 (a) NDVI monthly mean product map for the month of July 1985 for the African continent (coloured single-band image: bright green indicates healthy dense vegetation, black indicates no vegetation). (b) The vegetation dynamics over a year for the Continent of Africa: the monthly mean NDVI scenes for January (in red) May (in green) and September (in blue) are superimposed on one another in a false colour composite. Courtesy Remote Sensing Unit, University of Bristol; D. Lloyd and G. D'Souza. Original data from NOAA 1986.

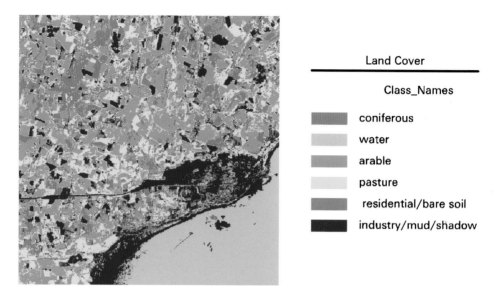

Land Cover

Class_Names

coniferous
water
arable
pasture
residential/bare soil
industry/mud/shadow

Plate 4.5 Land-cover map of area between Folkestone and Ashford, Kent, UK. The Landsat TM image for May 1989 has been classified to make a thematic map of land cover. Original data from ESA 1989, distributed by Eurimage NTSC.

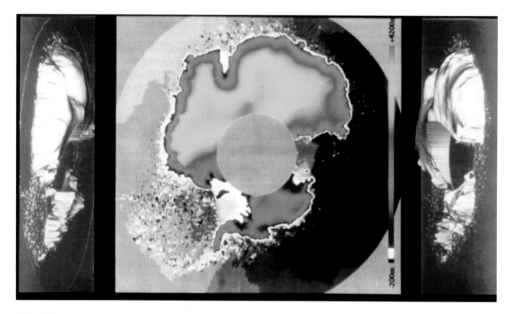

Plate 4.6 Topographic map of Antarctica. Vertical height data can be extracted from both the SAR imagery by the process of interferometry and by using data from the radar altimeter. The topographic map and computer-generated models shown have been derived from the altimeter data. Image courtesy of ESA and Eurimage: originally published n 'ERS-1-500 days in orbit' by ESA. copyright © ESA 1991, 1992 original data distributed by Eurimage.

variations if the emissivities of the different surfaces are similar. Many natural substances have emissivities of around 0.9 but metals that may be used for buildings are considerably lower. Thus, if a similar grey tone is obtained for a metal roof and for water on a thermal image, the roof is in fact warmer than the water. Although a grey-tone thermal image allows relative differences in emittance to be evaluated, it is possible to calibrate the image in such a way that absolute temperature differences may be determined. One way of doing this is by splitting the thermal infrared radiation by filters into two separate wavebands, which are measured alternately by the detector. A variable wedge is inserted behind one of the filters until the signal from both wavebands is similar. The position of the wedge at which the signals are equal is a measure of the temperature of the source because the wedge had earlier been calibrated by means of sources of known temperatures. An alternative method of obtaining absolute temperatures is by inserting sources of known temperature into the path of the scanning mirror so that, as the scanning mirror obtains a line of data, it initially sweeps across a known temperature source (which produces a specific signal) and at the end of the scan it sweeps across the other source. The temperatures of the sources are different and generally encompass the range of temperatures that one would expect to find for any survey.

Applications of Thermal Imaging

Thermal imaging has a number of applications in a range of disciplines.

1   Thermal imaging of urban environments allows heat loss surveys to be conducted. These images are usually colour coded (from blue for cold to white for hot) and indicate poorly insulated sites and locations where heat loss is excessive.

2   Leaks from underground pipes carrying hot water can undermine foundations and also result in substantial energy losses. Thermal scanning is a non-destructive technique for finding the sites of such leaks quickly.

3   In industrial situations, such as power stations or chemical installations, the failure of a containing vessel may release a dangerous gas or liquid into the environment. If the fluid is warm, the walls and seams of the vessel may be thermally examined in order to ascertain whether there are any weak spots.

4   Hand-held imaging scanners are used by the emergency services to locate people beneath rubble in disaster situations such as collapsed buildings. The detectors sense the heat from living persons and the rescue teams can then concentrate their efforts on saving those who are still alive.

5   Thermal imaging is used for night-time surveillance in order to study nocturnal animal behaviour or to detect the presence of intruders.

6   In geological investigations, two rock bodies may have similar reflectance properties and may not be distinguishable on a visible or near infrared image. However, if the bodies have very different thermal properties they would be differentiated on a thermal image.

7   Water is essential for agriculture and the presence of water along a fracture may produce an anomalous cold signature that can be detected by thermal means.

8   An extensive smoke cover in the visible spectrum often hides the sources of forest fires. However, a thermal image may reveal the sources (which are indicated by bright responses) through the smoke. Figure 3.8 shows a thermal image of Horseshoe, Arizona, taken in 1995 by an airborne Thermal Infrared Multispectral Scanner (TIMS). The pre-dawn image is dominated by mid-grey tonal variations though several bright signatures can be observed. These represent the sources of forest fires and such images allow fire-fighting efforts to be concentrated at the sources.

9   Heated water, which is often discharged into estuaries by power-generation plants, is a pollutant that may adversely affect the plant and fish life in the vicinity. Thermal imaging techniques can ensure that the heated water being discharged is within specified guidelines and that water movements are adequately dissipating the heat.

Figure 3.8 Forest fires displayed on airborne thermal imagery in the Horseshoe region of Arizona that allows the source to be pinpointed. Courtesy of NASA.

10 Volcanic monitoring can be carried out to help determine whether an eruption is likely as hot magma moves from depth towards the surface. The appearance of a thermal anomaly in conjunction with other signs such as gravity changes and bulging of the ground may be a precursor to a full eruption.

A night-time thermal image of part of Washington DC is shown in Figure 3.9a. Water bodies such as the Washington Channel (1, Figure 3.9b) and the Reflecting Pool (2) have a characteristic warm signature on a night-time thermal image. The major routes such as Pennsylvania Avenue (3) also yield a bright signature, unlike the Francis Case Memorial Bridge (4) and the Dwight D. Eisenhower Freeway (5) which are made of materials with different thermal properties. The railway system (6) produces a dark, cold signature. The Mall (7) is associated with two tones of intermediate grey, the inner darker one

being grass while the outer one is trees. Many buildings do not appear to be emitting much heat and consequently appear black on the image. Examples include the Arts and Industries Building (8), the National Air and Space Museum (9) and the National Museum of Natural History (10). However, other buildings such as the US Capitol (11) and the National Gallery of Art (12) are associated with slightly higher emittances and produce a dark grey signature.

Airborne Hyperspectral Imaging

Most airborne remote sensing systems obtain data in relatively few bands (approximately 4–6), with typical spectral resolutions of 1–2 µm (1,000–2,000 nm). Airborne hyperspectral systems, however, obtain data in a large number of wavebands with extremely high spectral resolutions. Various types of hyperspectral system are currently in operation, though the Airborne Visible Infrared Imaging Spectrometer (AVIRIS) flown by NASA is probably the most widely known one. AVIRIS acquires data in 224 bands in the 0.4–2.5 µm range. Each individual band covers a 10 nm range. AVIRIS data may be presented as black and white images for individual bands. However, it is also possible to display the data in a non-image format in which the continuous spectra for each individual pixel are presented as a graphical output (Figure 3.10). The absorption spectra for different substances are different but sensors that image wide wavebands, because the absorption bands are often very narrow, may not detect these differences. Thus, identifying specific features using broad-band sensors such as Landsat may not be possible. The very high spectral resolution obtained by hyperspectral systems allows detailed spectral information to be obtained and may permit the surface material to be identified. An image produced from AVIRIS data for the Cuprite region of Nevada is included on the WWW site accompanying this book.

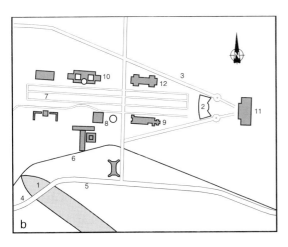

Figure 3.9 (a) Night-time thermal image of part of Washington DC, courtesy of Battelle. (b) Interpretation of image (see main text for numbered locations).

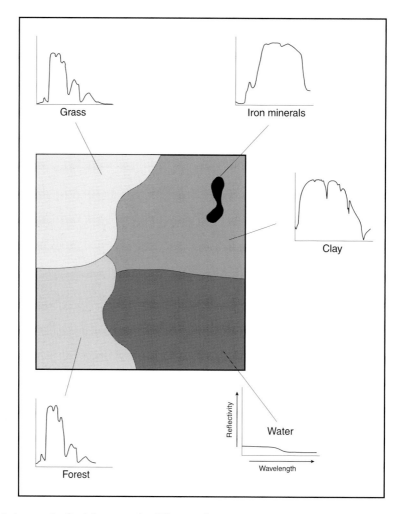

Figure 3.10 Typical spectral reflectivity curves for different substances. Hyperspectral imaging in very narrow wavebands allows the production of such curves and the possible identification of the features present in an image.

Multisensor Aerial Surveys

More information can be gathered by obtaining data for the same area by means of different imaging systems. The United States Department of Energy has conducted the Airborne Multisensor Pod System (AMPS) programme. This involved the imaging of selected targets by a number of instruments that are designed to be accommodated in pods carried by a Lockheed RP-3A aircraft. Pod 1 contains synthetic aperture radar (see section 3.2) and pod 2 carries a range of sensors that include:

1 a Daedalus Airborne Multispectral Scanner (AMS) which can obtain data in ten wavebands (though data cannot be obtained for more than six bands at the same time);

2 a Wild Heerbrugg RC-30 large-format camera (LFC) which obtains high-resolution stereo photographs;

3 a Barr and Stroud thermal imager;
4 a Compact Airborne Spectrometer Imager (CASI), which is a hyperspectral pushbroom system that obtains up to 288 spectral bands in the 0.4–0.9 μm range with a spectral resolution of up to 1.8 nm;
5 a video camera and a low-light solid-state CCD imager.

The evaluation of the use of these sensors for discriminating a range of targets which encompass urban, agricultural and coastal sites will allow the development of sensors that may eventually be incorporated in satellite remote sensing systems.

A Multispectral Electro-optical Imaging Scanner system (MEIS) which obtains digital data for eight wavebands in the visible and near infrared has produced airborne multispectral images in Canada. Current images have a resolution of 25 cm. An example of MEIS imagery, which can be used as an image-interpretation exercise, is included on the WWW site that accompanies this book.

3.2 IMAGES OBTAINED FROM LOW-ORBITING SATELLITE PLATFORMS

Landsat System

On 23 July 1972, the first civilian satellite dedicated to obtaining repetitive remote sensing data of the Earth with a spatial resolution of less than 100 m was launched. Terminology concerning the naming of satellites and the wavebands that they sense can be confusing. This satellite was originally called the Earth Resources Technology Satellite 1 (ERTS 1) but is now referred to as Landsat 1. So far, two series of Landsats have been launched: Landsats 1–3 and Landsats 4–5. (Before satellites are launched, they are often designated by letters. For example, the satellite termed Landsat A before launch became Landsat 1 once it was in orbit.) Satellite systems have a number of advantages and disadvantages compared with aerial remote sensing. A single Landsat scene covering an area of 185 × 185 km is the equivalent of around

1,000 aerial photographs. To form all these photographs – which may have been taken under varying lighting conditions – into a mosaic is impracticable. However, satellite images allow large areas that have been imaged in a very short time period to be analysed synoptically. Scale distortions are minimised because topographic variations are very low compared with the altitude at which the satellite revolves, and only a small rotation in the mirror angle is required to image a wide area.

Landsats 1–3

The first Landsat series consisted of three satellites that were launched between 1972 and 1978 into near polar sun-synchronous orbits at a nominal altitude of 918 km. This orbit prevented images from being obtained at very high latitudes (above 81 °N or 81 °S). The satellites obtained their data for a 185 km wide swath at about 9.30 a.m. local time while travelling in a southward direction. The orbital plane for a Landsat satellite is approximately constant but, as the Earth rotates beneath it, different strips of the Earth are imaged (Figure 3.11). A combination of the speed of the satellite and the rate of the Earth's rotation results in 14 such swaths being obtained in one day which, at the equator, are thousands of kilometres apart. A continual daily shifting of the area imaged results in total global coverage every 18 days, after which the cycle is repeated. However, by timing the launches properly and having a number of satellites operating simultaneously, images of the same area can be obtained at shorter time intervals. For example, Landsat 1 continued to operate until 1978 and Landsat 2 was operational from 1975, so that for a three-year period, two Landsat satellites were obtaining data of the Earth.

The early Landsat satellites carried two types of imaging system: a return-beam vidicon (RBV) and a multispectral scanner (MSS) (Table 3.1). The 3-band RBV with an 80 m resolution did not prove successful and was replaced on Landsat 3 by a single-band (0.5 μm–0.75 μm) 40 m resolution system. The RBV system has not been used in the second Landsat series. The MSS system is a transverse scanner (Figure 2.26a)

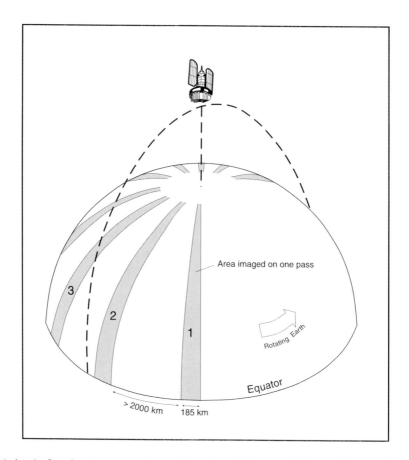

Figure 3.11 Orbital path of Landsat: 1 represents the area imaged on the first pass. Adjacent paths imaged on the same day (2 and 3) are separated by over 2000 km at the equator. The 'gaps' are filled in by imagery obtained during the following days.

that obtains data in four wavebands corresponding to the green (band 4), red (band 5) and infrared (bands 6 and 7) (Table 3.1).

The MSS onboard Landsat has proved a great success and has resulted in the acquisition of hundreds of thousands of datasets for the Earth. These datasets provide an invaluable resource for evaluating the effects of natural and human-induced changes in our environment. The scanning mirror rotates only 11.56 degrees to obtain data for a 185 km wide strip. The Landsat MSS does not collect its data line by line but for six lines simultaneously. If only one line of data was obtained for each sweep of the mirror, because of the speed of the satellite, the mirror velocity would have to be much greater. This would reduce the dwell time

and degrade the signal-to-noise ratio. The MSS system contains 24 detectors (six for each of the four bands). The IFOV for the Landsat MSS is usually given as 79 × 79 m (though this may change somewhat because of altitudinal variations in the satellite's orbit). The continuous signal is split into the four sensed bands and electronically sampled every 9 microseconds. This sample time is equivalent to a ground distance of approximately 56 m. Thus the ground-sampling cell for the MSS system is 79 m (along track) by 56 m (across track). Four digital numbers are assigned to each pixel depending on its reflectance in each of the four sensed bands and the data transmitted to receiving stations. When the onboard tape recorders were out of range of a receiving station, they retained the

Table 3.1: Landsat MSS and TM spectral bands

Band number	Landsat 1 (μm)	Landsat 2	Landsat 3
1 (RBV)	0.475–0.575	0.475–0.575	
2 (RBV)	0.58–0.68	0.58–0.68	0.5–0.75
3 (RBV)	0.69–0.83	0.69–0.83	
4 (MSS)	0.5–0.6	0.5–0.6	0.5–0.6
5 (MSS)	0.6–0.7	0.6–0.7	0.6–0.7
6 (MSS)	0.7–0.8	0.7–0.8	0.7–0.8
7 (MSS)	0.8–1.1	0.8–1.1	0.8–1.1
8 (MSS)			10.4–12.6

Landsats 4 and 5

TM band	Waveband (μm)	Spectral region
1	0.45–0.52	Blue
2	0.52–0.60	Green
3	0.63–0.69	Red
4	0.76–0.90	Near infrared
5	1.55–1.75	Near infrared
6	10.4–12.5	Thermal infrared
7	2.08–2.35	Near infrared

data until it was possible to transmit them. The data are stored on computer-compatible tapes (CCTs) for areas of approximately 185 × 185 km. Each of these scenes contains approximately 3,240 pixels (across track) by 2,340 lines (along track) and is given a unique path and row number for identification purposes.

Landsats 4–5

In 1982, the first of the second generation of Landsat satellites (Landsat 4) was put into orbit. The orbits of Landsats 4 and 5 are also near polar and sun-synchronous but they are at a lower altitude than the previous series (approximately 705 km). The satellites have a higher velocity on account of their lower altitude and consequently have a shorter repeat cycle, of 16 days.

Landsats 4 and 5, as well as retaining the MSS system, also carry a more sophisticated data-acquisition system termed the Thematic Mapper (TM). Landsats 4 and 5 did not carry onboard recorders. Instead the data were transmitted to a geostationary Tracking and Data Relay Satellite from which they were transmitted to the ground.

Some confusion may arise when the wavebands sensed by the MSS and TM systems are being discussed. MSS bands 4, 5, 6 and 7 on Landsats 1, 2 and 3 are equivalent to MSS bands 1, 2, 3 and 4 on Landsats 4 and 5. Thus, if one refers to band 4 on Landsat 1, this is the green band, whereas on Landsat 5 this is a photographic infrared band. An examination of Table 3.1 also shows that TM band 7 is at a lower wavelength than TM band 6. This came about because, when the TM system was initially designed, there were no plans to obtain data in the TM 7 waveband. However, this is an important atmospheric window for geologists and this band was included after the others had been assigned their numbering sequence.

There are a number of differences between the TM and MSS systems:

1 TM data are obtained by the scanning mirror sweeping eastward and westward but MSS data are only obtained on eastward sweeps.

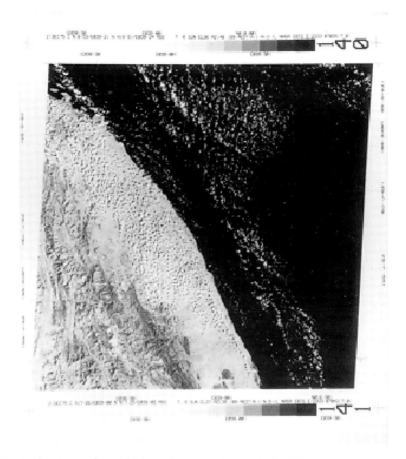

Figure 3.12 MSS 7 Landsat image of part of Eritrea. Area approximately 185 x 185 km.

2 The Instantaneous Field of View (IFOV) for the TM system for bands 1–5 and band 7 is 30 m whereas the MSS has an IFOV of 79 m. The IFOV for TM band 6 (thermal band) is 120 m.

3 The incoming radiation in the TM system is directed by means of a telescope to one of two focal planes depending on the waveband that is being sensed. The sensors for the TM bands contain 16 detectors except for TM 6, which has 4 detectors the TM system therefore has 100 detectors while the MSS system has 24.

4 The radiometric resolution for MSS and TM is different. The data for the MSS system are recorded in a 6-bit format (0–63), though after transmission to their ground receiving station, data for MSS bands 4, 5 and 6 are decompressed to 7 bits (0–127). TM data are obtained with an enhanced radiometric resolution of 8 bits (0–255).

5 The scan angle for TM is slightly greater (15.4 degrees) than that of the MSS system.

A TM scene is formed of over 6,000 pixels (across track) by about 5,660 lines (along track). The higher spatial and radiometric resolution and increased number of bands sensed for the TM system compared with MSS means a much greater amount of data has to be transmitted to Earth for the former. The transfer rate for these data is 85 Megabits per second for TM and 15 Megabits per second for MSS.

A standard false colour composite is produced from Landsat MSS data by projecting MSS 7 in red, MSS 5 in green and MSS 4 in blue. The larger number of bands for the TM system allows the production of a much greater number of false colour composites. A combination of TM 1, TM 2 and TM 3 projected in blue, green and red respectively forms an approximately natural colour image. The MSS system does not obtain any data in the blue range though band 1 in the TM system extends into the blue part of the spectrum. (Scattering affects the shorter waveband much more than the longer wavebands and the signal-to-noise ratio for MSS was such that a meaningful signal could not be assured. The better signal-to-noise ratio for the TM system allows these shorter wavelengths to be sensed though the effect of atmospheric scattering is still noticeable on the images.) TM 1 is particularly suited for penetrating water as blue light travels farther through a water column than longer wavelength radiation. TM 2 is equivalent to the green reflectance band for vegetation in the vis-

ible range. TM 3 and TM 4 straddle the red edge of vegetation (TM 3 below it and TM 4 above it). Vegetation indices can be constructed from reflectance values obtained by these bands, which allow estimates of vegetation density to be made. TM 5 tends to give 'sharp' images and is not affected by atmospheric scattering to any great extent and can also give an indication of the moisture content of vegetation. TM 7 was introduced at the request of the geological community because hydroxyl ions in clays and other minerals located around alteration zones (which may occur near ore bodies) have prominent absorption bands which fall within this waveband. TM 6 is within the thermal part of the infrared; features with similar reflectance characteristics but different thermal properties may thus be differentiated.

The Landsat series of satellites has been the most successful to date, providing worldwide coverage for over 27 years. However, there have been failures. Landsat 4 did not function properly because the solar panels and data handling systems failed. The greatest loss (to date) for the Landsat series was the disappearance of Landsat 6 soon after its launch in

Figure 3.13 MSS 4 (a) and MSS 7 (b) image of the Sperrin Mountains, Northern Ireland. Scattering is much more prevalent on the short-wavelength image (MSS 4) compared to the longer near infrared image (MSS 7). Area approximately 35 × 35 km.

1993. This was to have had capabilities similar to those of Landsat 5 but included an enhanced Thematic Mapper which was to obtain panchromatic single-band data (0.5–0.9 μm) with a resolution of 13 × 15 m. Landsat 7, which was successfully launched in April 1999, also carried the enhanced Thematic Mapper (ETM+) and can obtain thermal data with a 60 m spatial resolution.

Landsat Images

A standard single-band MSS 7 image of Eritrea and part of the Red Sea is shown in Figure 3.12. The land–sea boundary is clearly delineated by a sharp difference in tone: the water is black because of the very low reflectivity in the near infrared. Scattered clouds appear white; some can be seen over the coastal zone. On land, the vegetation cover is so poor that bare rock is exposed at the surface. The area covered by the image is approximately 185 × 185 km though it is parallelogram-shaped rather than square. This is because in the time between Landsat scanning its first and last lines of data for a 185 km long image (approximately 25 seconds), the Earth has rotated eastward beneath the path of the satellite. Thus the area image on the ground is not square but rhombus-shaped. Information accompanying the image provides details on the date of acquisition, the co-ordinates of the centre of the scene, the illumination angle and direction, the band number, the receiving station, the satellite which obtained the data, and a scene identifier number.

The effects of scattering on Landsat data are well illustrated by comparing an MSS 4 and MSS 7 image

Figure 3.14 The four MSS bands for the same region. Clockwise from top left: MSS 4; MSS 5; MSS 7 and MSS 6. The high reflectance for the vegetation in MSS 6 and MSS 7 is very evident.

of the same region (Figure 3.13). The lineaments of the ridges and valleys are much more clearly defined on MSS 7 where the effect of atmospheric scattering is minimal compared with MSS 4, which collects data at a shorter wavelength. Figure 2.11 illustrated that the reflectance for vegetation increases greatly in the infrared part of the electromagnetic spectrum compared with the visible range. The four MSS bands for a vegetated area are shown in Figure 3.14. The MSS 6 and MSS 7 images (infrared) are brighter (i.e. have

a much greater reflectance) than either MSS 4 (green) or MSS 5 (red). Note also the very low reflectance of the ocean in the infrared range along the northern edge of the images.

Figure 3.15 shows all seven TM bands for a 15 × 15 km area in northeast Ireland. The effects of a thin cloud cover over the north central part of the area are most noticeable on the lower wavelength bands (TM 1, TM 2 and TM 3) compared with the images taken at longer wavelengths (TM 4, TM 5 and TM 7). The

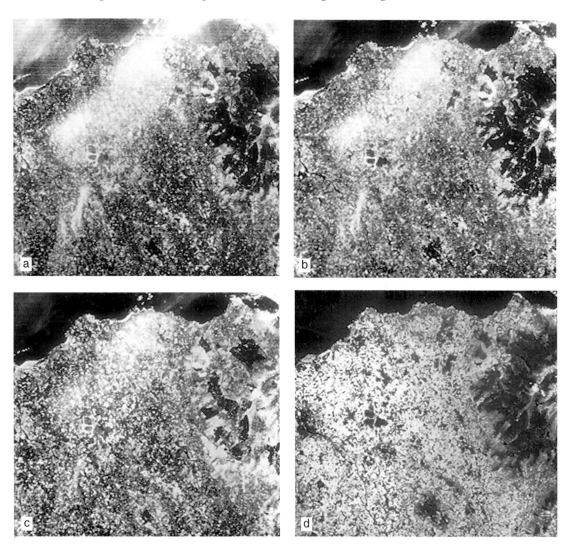

Figure 3.15 (a-d) Seven TM bands for a vegetated area in temperate terrain (northeast Ireland): (a) TM 1; (b) TM 2; (c) TM 3; (d) TM 4.

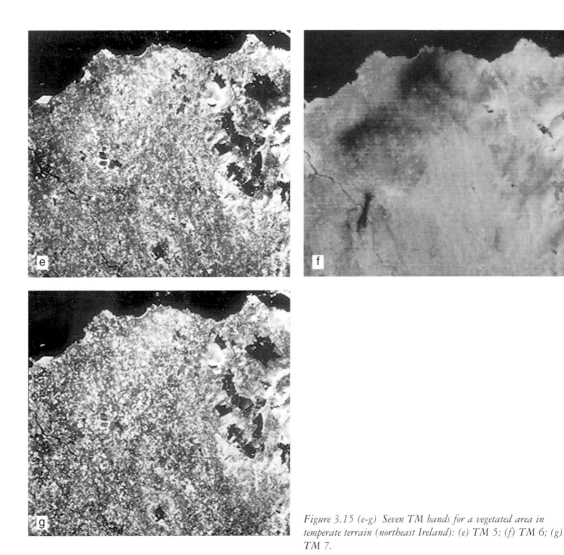

Figure 3.15 (e-g) Seven TM bands for a vegetated area in temperate terrain (northeast Ireland): (e) TM 5; (f) TM 6; (g) TM 7.

cloud is characterised by a high reflectance in the visible range, though it appears dark on the thermal band (TM 6), indicating cooler temperatures. The high reflectance for TM 4 is indicative of vegetation. Coniferous plantations are characterised by dark signatures and regular outlines on TM 4 though the sharp outlines of the forest are best seen on the TM 3, TM 5 and TM 7 images. The forests are less obvious on the thermal image but they also appear slightly darker than the cultivated regions. The largest town in the area, Coleraine (C), is best seen on TM 4, where

it has a low reflectance, and TM 6, where it is associated with a higher thermal return. A bright diamond-shaped region on the coast (1, Figure 3.15h), an area of sand dunes, is less prominent on TM 1 and TM 4 than on TM 2, TM 3, TM 5 and TM 7. A subtle pale lineament (shown dashed on Figure 3.15h) can be observed on the thermal image but cannot be seen on the other images and does not correlate with any known natural or man-made feature. Care must always be exercised when interpreting satellite images. The dark signature observed on TM 4 (2,

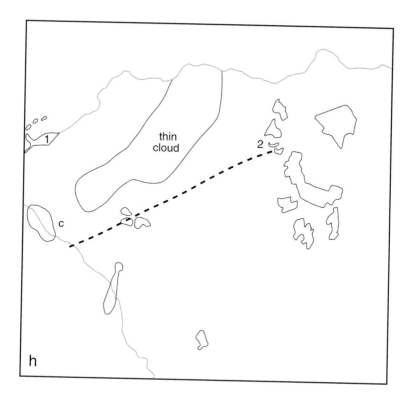

Figure 3.15 (h) Seven TM bands for a vegetated area in temperate terrain (northeast Ireland): (h) interpretation of imagery. C=Coleraine. Area approximately 41 × 41 km.

Figure 3.15h) is not a lake or forest but is the shadow of a cloud. The cloud itself has a high reflectance and can be seen to the southeast of the shadow.

Three TM bands may be combined in order to create a colour image. An approximately natural colour image is produced by projecting TM 1 in blue, TM 2 in green and TM 3 in red (Plate 3.1a). The image shows an agricultural region in which the vegetation is mainly green. Thin white clouds cover the western half of the image. The colour variations for the lake (from black to pale purple) indicate different water depths, in which the black areas represent the deepest parts. The small lakes at the top of the image are sufficiently deep for visible radiation not to be able to penetrate to the lake bottom; they thus have a black signature. There is no standard format for creating false colour composites from TM data. Particular band combinations are employed for specific tasks.

For example, if one were interested in investigating geological alteration zones, TM 7, which obtains data in the waveband where hydroxyls have important absorption bands, could be employed. Plate 3.1b shows bands 4, 5 and 3 projected in red, green and blue respectively. The longer wavelengths for these bands allow a more effective penetration of the thin cloud. On this image, reds and oranges represent crops while many of the blue areas represent bare soil. Image interpretation exercises using single-band and false colour TM images are included on the WWW site.

Système Pour l'Observation de la Terre (SPOT) System

In 1986, France launched the remote sensing satellite SPOT from French Guinea in Africa. To date, three

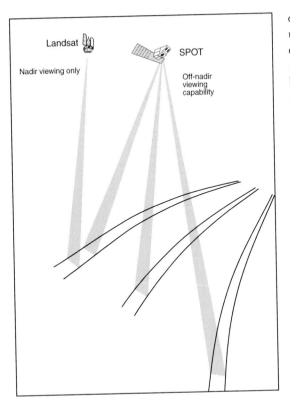

Figure 3.16 Viewing capabilities of Landsat and SPOT. Landsat can obtain data only from a swath directly beneath the satellite whereas SPOT has an off-nadir viewing capability.

ent angles (up to a maximum of 27 degrees) can be used to obtain data at distances of up to 450 km on either side of the satellite's groundtrack (Figure 3.16). This off-nadir viewing capability, not shared by the Landsat series, has a number of major advantages. The nominal repeat cycle for SPOT is 26 days but because it has the capability to look either to the east or the west of its groundtrack, a much quicker repeat cycle can be attained, the actual time interval depending on the latitude of the scene. At mid-latitudes (i.e. Seattle in the USA or Paris in Europe), a scene may be imaged up to 11 times in one nominal cycle. This versatility means that the satellite may be directed to monitor unpredictable events like forest fires or floods. Again, if cloud cover prevents images being obtained at one location, the viewing angle may be altered in order to collect data from a cloud-free region. The ability of SPOT to image the same scene from different viewpoints is analogous to an aircraft obtaining a set of overlapping aerial photographs. Stereoscopic viewing of a SPOT scene is therefore possible. (The overlap between adjacent Landsat images does allow some scope for stereoscopic viewing. However, the area of most overlap, high latitudes, is characterised by very small base-to-height ratios.) SPOT was specifically designed with the capacity to obtain world-wide stereo images and the base-to-height ratios are significantly greater than for Landsat.

SPOT is a pushbroom system that can obtain its data in two modes (Table 3.2). In multispectral mode (XS) SPOT 1, 2 and 3 obtain data at a radiometric resolution of 8 bits for three bands using 3,000 separate detectors. The resolution for this configuration is 20 m. SPOT 1, 2 and 3 may also obtain a single band panchromatic (P) image using 6,000 detectors in the 0.51–0.73 mm waveband with a 10 m resolution. SPOT, like Landsat, collects its data in the morning, crossing the equator at approximately 10.30 a.m. local time. The multispectral capabilities of SPOT 4 are similar to those of the earlier satellites but it also obtains data in a fourth waveband at 20 m spatial resolution and also carries onboard a VEGETATION instrument which images a wide swath (2,200 km) with a

other satellites in this series have been launched: SPOT 2 in 1990, SPOT 3 in 1993 and SPOT 4 in 1998. The orbital characteristics of the SPOT satellites are similar to those of Landsat but their data-acquisition systems are radically different. SPOT has an 832 km high near polar sun-synchronous orbit. The satellite does not use a scanning mirror to obtain its data but has a pushbroom system that uses charge-coupled devices. The data-acquisition system for SPOT consists of two High Resolution Visible (HRV) sensors, each of which images a 60 km wide swath immediately beneath the satellite. A 3 km overlap between the areas sensed results in a 117 km wide swath compared with the 185 km for Landsat. Although SPOT does not employ a scanning mirror, an onboard mirror that can be rotated through differ-

Figure 3.17 A 10 m resolution panchromatic SPOT image showing regular urban street patterns in New York. Copyright CNES 1995, courtesy of SPOT IMAGE.

1.1 km spatial resolution in the same three wavebands as the earlier SPOT satellites. The panchromatic band on SPOT 4 also has a narrower spectral resolution than the earlier satellites.

A multispectral SPOT image taken in 1992 of part of the coast of the Caspian Sea near Gurjev is shown in Plate 3.2. The high spatial resolution combined with the multispectral capabilities allows the movement of terrestrial sediments into the sea and around the coast to be monitored. A panchromatic 10 m resolution SPOT image of part of New York is shown in Figure 3.17. The dark rectangular region in the top centre is part of Central Park. The regular grid layout of streets can be discerned at this scale; such images are particularly useful for urban studies.

NOAA Meteorological Satellites

The National Oceanic and Atmospheric Administration (NOAA) launched a series of polar-orbiting satellites in 1970 (starting with NOAA-1) whose function was primarily to provide meteorological data. The initial satellite carried two scanning radiometers obtaining data in the 0.5–1.0 μm waveband with 4 km resolution and in the thermal band (10.5–12.5 μm) at 8 km resolution. In addition, two advanced vidicon camera systems obtained data at 0.45–0.65 μm with a resolution of 2.2 km. Later satellites, beginning in 1978 with the TIROS-N series (Television Infrared Operational Satellite), carried more sophisticated instruments. This series

Table 3.2 Spectral bands for the SPOT satellites

Multispectral mode (XS)

Band 1	0.50–0.59 μm (green)
Band 2	0.61–0.68 μm (red)
Band 3	0.79–0.89 μm (near infrared)

Band 4 (SPOT 4 only) 1.58–1.75 μm (near infrared)

Panchromatic mode (P)

0.51–0.73 μm (SPOT 1, 2 and 3)
0.61–0.68 μm (SPOT 4)

Table 3.3 NOAA band designation

Band number	Waveband (μm)	Spectral region
1	0.58–0.68	Red
2	0.72–1.1	Near infrared
3	3.55–3.93	Thermal infrared
4	10.3–11.3	Thermal infrared
5	11.5–12.5	Thermal infrared

operates in a near polar sun-synchronous orbit at an altitude of 833 km. Images are obtained at least twice a day in the visible range. The main imaging sensor is the Advanced Very High Resolution Radiometer (AVHRR) which obtains a 2,400 km wide swath with a nadir resolution of 1.1 km in five wavebands (Table 3.3).

Many of the latest generation of satellites also carry the TIROS Operational Vertical Sounder (TOVS), which consists of:

1 a High Resolution Infrared Radiation Sounder (HIRS);
2 a Stratospheric Sounding Unit (SSU);
3 a four-band Microwave Sounder Unit (MSU).

(NOAA-15 has onboard advanced versions of the SSU and MSU and has an additional sixth band centred at 1.6 μm.) Data from the NOAA satellites are collated and transmitted in a variety of formats. Local Area Coverage (LAC) imagery is at full resolution (1.1 km) whereas Global Area Coverage (GAC) represents a resampled data set with a 4 km resolution. Both datasets are stored onboard before transmission but High Resolution Picture Transmission (HRPT) transmits the data to ground-based recording stations as they are obtained.

Apart from directly providing meteorological information about temperature, cloud cover and humidity variations, the TIROS datasets, after suitable processing, also provide other important indirect information relevant to the environment. Biomass burning, apart from the temporary loss of land (but often permanent loss of habitat), releases a large amount of material into the atmosphere which needs to be taken account of in

the monitoring of climate change. A monthly fire index atlas has been constructed for the entire African continent from data obtained from thousands of TIROS images. Changes in the fire index have been correlated with movements of the dry season through the continent. (See Arino and Melinotte (1995) for a more detailed discussion.) Continental vegetation indices can be produced from particular bands on the NOAA satellites. Plate 1.2b shows part of an AVHRR image of the British Isles and examples of NOAA images are illustrated in Chapter 4.

Defence Meteorological Satellite Program (DMSP)

The United States' military operates an independent meteorological satellite series called the Defence Meteorological Satellite Program. The sun-synchronous near polar orbiting satellites operate at an altitude of 830 km and map cloud distribution and temperatures twice daily. The operational linescan system (OLS) obtains data with a radiometric resolution of 6 bits in the 0.47–0.95 μm waveband and also acquires data with 8-bit resolution in the 10–13.4 μm waveband. The latter system is calibrated to measure temperatures between 190 °K and 310 °K. Data are obtained with a nadir resolution of 0.55 km for a 3,000 km wide swath. The satellites have the capability to obtain data, mainly within the visible range, at night by using a Photo-Multiplier Tube (PMT) in the imaging system. Resolution at night is approximately 2.7 km at nadir. Night-time images dramatically show the position of urban centres and often the main routeways connecting cities. This 'light pollution' has now reached such a level that astronomers are expressing concern about their inability to view low-illumination objects in space because the signals are being swamped by terrestrial light sources. A night-time image of the United States is shown in Figure 3.18. The major cities such as Chicago, New York, Dallas and Los Angeles are clearly visible, as are many of the smaller towns situated along the main routes which connect the major cities. A marked east–west divide can be observed on the image delineated by the 100 degree west line of longitude. To the

Figure 3.18 DMSP night-time image of the United States of America showing the urban centres as bright spots produced by upwelling light. Image courtesy of NSIDC.

west of this line there is a decrease in the amount of light emanating from the USA, reflecting a lower population concentration.

The satellites also carry on board three other instruments with which various aspects of the atmosphere and the Earth's surface can be measured. A Special Sensor Microwave Imager (SSM/I) is a radiometer designed to measure electromagnetic radiation in the microwave region emitted from the land, ocean and atmosphere at four wavelengths (0.35 cm, 0.8 cm, 1.35 cm and 1.55 cm). Data from a Special Sensor Microwave Temperature Sounder (SSM/T-1) are obtained in seven channels between 0.5 cm and 0.6 cm and can be used to reconstruct a temperature profile for the lower stratosphere and troposphere. A Microwave Water Vapour Sounder (SSM/T-2) is a five- channel device which obtains data at 0.2 cm, 0.3 cm and in three channels centred near 1.64 mm.

Japanese Earth Resources Satellite

The Japanese Earth Resources Satellite (JERS-1) was launched on 11 February 1992. It is in a near polar sun-synchronous orbit at a height of 568 km and has a repeat cycle of 44 days. JERS-1 carries onboard an active imaging radar (see later section) but also

Table 3.4 JERS-1 band designation

Band number	Waveband (μm)	Spectral range
1	0.52–0.60	Green
2	0.63–0.69	Red
3	0.76–0.86	Near infrared
4	0.76–0.86	Near infrared
5	1.60–1.71	Near infrared
6	2.01–2.12	Near infrared
7	2.13–2.25	Near infrared
8	2.27–2.40	Near infrared

measures passive electromagnetic radiation in eight wavebands extending from the visible to the near infrared (Table 3.4). Bands 3 and 4 both obtain data for the same waveband, band 3 from directly beneath the satellite, while band 4 detectors are angled by 15.33 degrees to obtain data from in front of the satellite, allowing along-track stereoscopic viewing. JERS-1 is a pushbroom system which images a 75 km wide swath with a spatial resolution of 18 m $\times$ 24 m and a radiometric resolution of 6 bits (64 grey levels). The radar system obtains data at a 35 degree look angle for a 75 km wide swath at 18 m resolution using a wavelength of 23 cm (L band).

Imaging in the Microwave

Images can be obtained by remote sensing systems operating within the microwave part of the spectrum. However, as shown in Figure 2.3a, the amount of naturally emitted radiation at microwave wavelengths is very low. Therefore, in order to measure these weak signals, large areas are imaged, this results in relatively poor spatial resolutions. Choosing the wavelength at which the data are obtained allows either the surface or the atmosphere to be investigated. The variations in microwave signals emanating from a surface and measured by an antenna are generally expressed as temperature variations (similar to a thermal image) in which the 'temperature' is the equivalent temperature of a blackbody which would produce the same energy as the measured microwave signal. Passive microwaves can provide information on soil moisture, water vapour variations and temperature. An alternative approach to producing images by using natural microwaves is to irradiate the surface artificially with a source of microwaves and measure the returned signal. This approach allows the measurement of larger signals and consequently the spatial resolution of such active microwave imagers is . much better than for passive microwave imagers. Active imaging in the microwave is termed 'radar'. Radar systems are operated from airborne and spaceborne platforms. However, it is considered within this section because the full potential of radar imaging is only now being realised, partly as a result of the

wealth of information that spaceborne sensors have provided.

Radar: Theoretical Concepts and Systems

Radar (RAdio Detection And Ranging) is the most important active system that is applied in the field of remote sensing. It has a number of major advantages over passive systems:

1   Radar produces its own electromagnetic radiation with which it illuminates the target. It is thus not dependent on solar illumination and may be operated at night. Many passive satellite systems also obtain data when the solar illumination is coming from a particular direction, usually the southeast for the northern hemisphere. Features with particular trends may be subdued on these images. Radar may image the same area from different directions, thus increasing the likelihood that all the features within the scene will be recorded.

2   A great disadvantage of a satellite such as Landsat is that cloud cover may prevent data from being obtained for a particular area. Extensive and sustained cloud cover is particularly prevalent in tropical regions, so that a region may not be imaged for years by passive sensors. Electromagnetic radiation at wavelengths at which some radar systems operate can penetrate cloud and thus image the ground below.

An area of Indonesia imaged by Landsat and radar is shown in Figure 3.19. Very little information can be obtained from Landsat because the presence of cloud and the extensive forest cover prevent the acquisition of data on terrain characteristics. The radar image, however, is unaffected by clouds and in addition a number of lineaments and variations in signatures can be observed as a result of the presence of different rock types. Because it operates at much longer wavelengths than the visible and near infrared bands obtained by many satellite systems, radar can often provide detailed information on areas

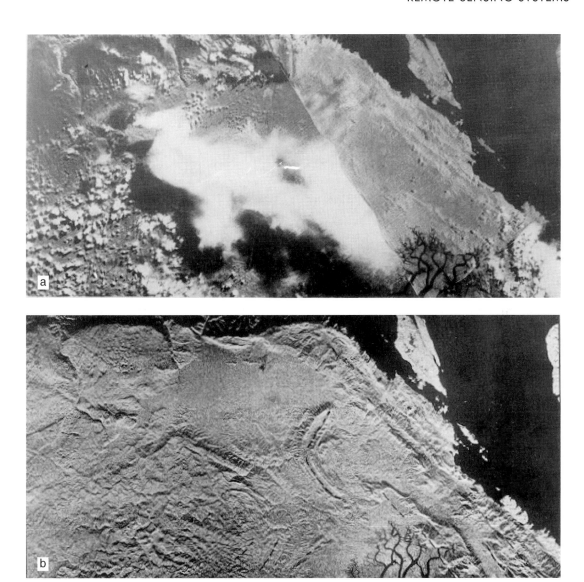

Figure 3.19 (a) MSS 7 image of part of Indonesia which provides little information because of cloud cover and forest cover. (b) Radar image of the same region acquired by the Space Shuttle Columbia in 1981 reveals much more information. Width 100 km. Image courtesy of NAS/JPL.

which appear uniform at lower wavelengths. Part of Columbia, imaged by Landsat, is shown in Figure 3.20a. On this image, the dark-toned western region represents a grassland area and the pale area to the east is a heavily forested area. Little information can be obtained from Landsat for the grassland but on radar the same region is seen to be dominated by extensive dendritic drainage networks (Figure 3.20b). Note, though, that the river flowing through the forest (pale regions on both images) can be detected on Landsat but is virtually undetectable on radar. In order to use remote sensing for the acquisition of

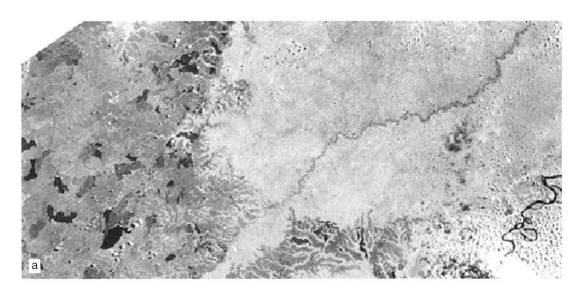

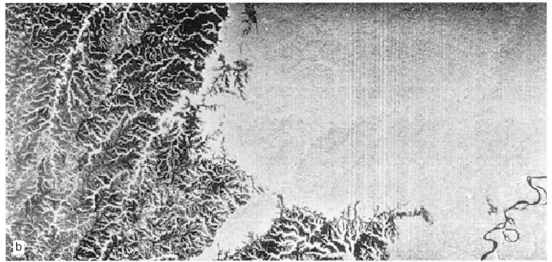

Figure 3.20 (a) MSS 7 image of Meta, Columbia, showing a pale forest area to the east and a darker grassland region in the west (b) Radar image of the same region acquired by the Space Shuttle Columbia in 1981 which shows the presence of an extensive dendritic drainage pattern in the grasslands. Width 100 km. Image courtesy of NASA/JPL.

information for any particular area, it is important to view the area on many different types of image at different wavelengths so that the maximum amount of information can be obtained. Conclusions reached solely on what can be observed on one image may not be justified.

Airborne radar was originally developed in order to obtain data about an area without the necessity of overflying the scene. Such Side Looking Airborne Radar (SLAR) systems emit a short-duration pulse of microwave electromagnetic radiation to the side of the aircraft. The part of the image nearest the aircraft is known as the near range and the part farthest from the sensor is termed the far range (Figure 3.21a). The

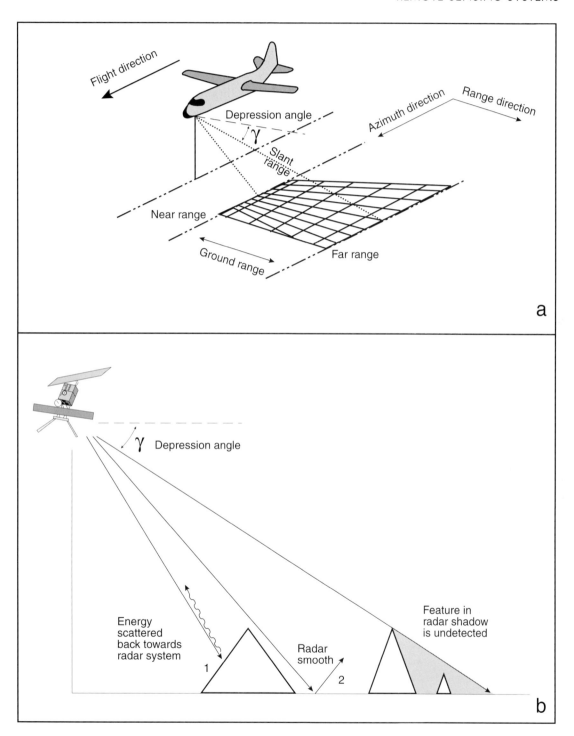

Figure 3.21 (a) Terminology used in radar imagery. (b) Variation in radar signature with strength of returned signal. 1 has a strong returned signal and will appear bright, whereas no signal is returned from surface 2 and it is shown dark. Features may remain undetected if they are in a radar shadow

depression angle (γ) for a radar system is the angle measured from the horizontal to the line joining the transmitting antenna and the object imaged. Thus the depression angle is greater in the near range than in the far range. Two other terms regarding angular relationships are common in radar literature: the look angle and the incidence angle. The look angle is the angle between the vertical and the line joining the antenna to the target. It is thus equal to 90 degrees minus the depression angle. The incidence angle is the angle between a line joining the target to the antenna and a line at right angles to the surface. For a flat surface and neglecting the curvature of the Earth, which is small for the distance imaged, the incidence angle is equal to the look angle. Data acquisition for a radar system differs from Landsat because images are not obtained directly below the platform and the antenna has a fixed orientation, unlike the Landsat scanning mirror. Radar images are built up in a series of narrow strips in the range direction by the forward movement of the airborne (or spaceborne) platform in the flight (azimuth) direction. In order to create a radar image, the system must record two significant parameters: time and the strength of the signal that is scattered back from the target (radar return).

When a radar signal intersects a target, it is reflected and scattered to a greater or lesser degree. A surface that scatters a large amount of the incident energy back towards the antenna has a high radar return (1, Figure 3.21b), and will be assigned a bright signature on the image. If the transmitted energy is mainly reflected (2, Figure 3.21b) then very little electromagnetic radiation will be backscattered and this target is assigned a dark signature. The transmitted and returned pulses of electromagnetic radiation travel at the speed of light. The radar systems measure the time lapse between transmitting and receiving the scattered pulse, which, because of the constant velocity of the wave, is a measure of how far the target is from the antenna. This distance is known as the slant range. A short time duration means the target is nearby (near range) whilst a longer time lag means the object is farther away. Thus a target can be correctly positioned and assigned a

signature on the radar image by measuring the strength and timing of a returned pulse.

Factors that Control Radar Signatures

The signatures obtained by radar systems are a function of the characteristics of the radar system such as wavelength, depression angle and polarisation, and also a function of the properties of the surface being imaged such as surface roughness, dielectric constant and aspect with respect to look direction. These characteristics and properties are discussed below.

The microwave range of the electromagnetic spectrum extends from 0.1 cm to 100 cm though commercial radar systems are designed to operate at a particular wavelength within a specified waveband designated by a letter code (Table 3.5). The letter codes were introduced when radar was being developed during the Second World War when the wavelengths being tested were military secrets.

Short-wavelength radars are used in meteorological studies and for weather forecasting. These radars may be ground-based and detect the scattering from rain droplets within clouds. The strength of the radar return is dependent on the amount of water within the clouds. In order for satellite-based systems to have all-weather capabilities they operate at longer -

Table 3.5 Nomenclature for waveband designation in the microwave

Waveband	Wavelength (cm)
Ka	0.8–1.1
K	1.1–1.7
Ku	1.7–2.4
X	2.4–3.8
C	3.8–7.5
S	7.5–15.0
L	15.0–30.0
P	30.0–100.0

Discrepancies between the limits of the wavebands may occur. Kondratyev et al. (1996) show the L band as extending from 15.8 to 63 cm and the Ka band from 0.72 to 1.0 cm. Modified from Sabins 1997.

Source: Modified from Sabins 1997.

Note: Discrepancies etc. 0.72 to 1.0 cm.

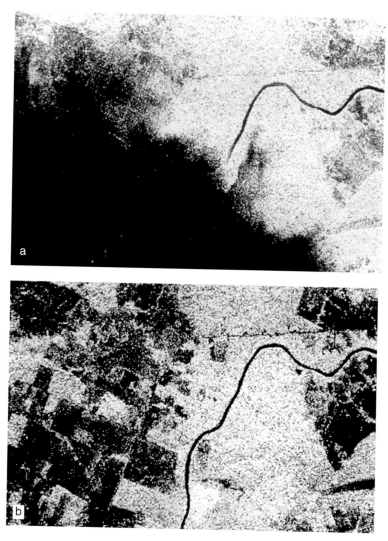

Figure 3.22 (a) X- and (b) L-band radar images of the Brazilian rainforest. The short wavelength X band cannot penetrate the rain cloud. Width of image 15 km. Image courtesy of NASA/JPL.

wave-lengths, often within the L band with a wave-length of approximately 23 cm. The presence of a heavy rainstorm in the Brazilian forest is shown by a low radar return on an X-band short-wavelength radar (Figure 3.22a). However, at a longer wave-length (L band) the radar can penetrate the rain and image the terrain which is invisible at the shorter wavelength (Figure 3.22b). The microwave radiation that is transmitted by radar systems is polarised with either a vertical or horizontal orientation. When this radiation intersects a target, a proportion of it can become depolarised by multiple reflections. The proportion that becomes depolarised will vary depending upon the original orientation of the transmitted wave and the nature of the surface. Thus layered rocks which form mainly horizontal planes will produce a different degree of depolarisation from that of a forest which consists of vertical trunks and horizontal branches. The antenna which measures the radar return may be designed to measure it in the same orientation as the transmitted wave or to measure the depolarised radiation whose vector is at right angles to the transmitted wave. Thus a radar system may be a vertical transmit – vertical return (VV); horizontal transmit – horizontal return (HH); vertical transmit – horizontal return (VH) or horizontal transmit – vertical return (HV) (Figure 3.23a). An HH and VV system is called a parallel-polarised system and a VH and HV system is cross-polarised. (Some radar systems transmit in one plane but record

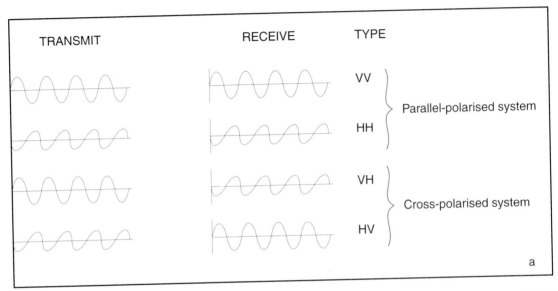

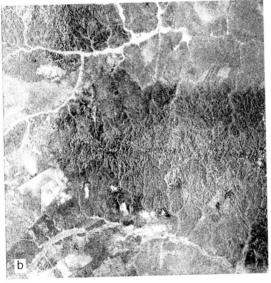

Figure 3.23 (a) Parallel and cross-polarised radar systems. (b) HH and (c) HV images of forestry in Indonesia obtained by the Space Shuttle. Width of image is 50 km. Courtesy of NASA/JPL.

in both.) The same target may produce different signatures depending on the particular combination of polarisations that are used. An area of Indonesia imaged by the Space Shuttle at a constant wavelength (L band) but in two polarisation modes is shown in Figure 3.23. In HH mode, the rivers produce a high radar return (bright) while the rest of the scene yields an intermediate (grey) radar return (Figure 3.23b). However, in HV mode the area is seen to consist of two very different signatures, a bright one surrounding a region that is delineated by straight edges which is associated with a dark signature (Figure 3.23c). The bright (high) return parts of the scene represent virgin rainforest whereas the dark areas show the locations where the natural forest has been cleared to produce managed plantations. The lack of trees in these regions results in little backscattered energy with a vertical orientation. The HV image is superior to the HH image for evaluating the extent of rainforest destruction in this area.

A feature in the far range is imaged at a smaller depression angle than if it were in the near range. For example, a mountain will have a longer radar shadow in the far range. In the far range, the mountain will be characterised by a bright signature for the side facing the sensor and a long black signature for the shadow. Features may be accentuated by this long shadow which might go undetected if they were in the near range. A disadvantage of the long shadow is that features which may be present in the shadow will go undetected (Figure 3.21b).

The trend of linear features such as valleys, fractures and ridges with respect to the look direction is an important factor in determining whether they are enhanced or suppressed. Features trending at a high angle (> 45 degrees) to the look direction tend to be highlighted, while those parallel to the look direction are subdued. Figure 3.24 shows a fracture which, if the illumination is from the west, produces a high radar return arising from a sloped face (CD) and a radar shadow (BC) which make it prominent. There is little contrast between the fracture and the background when the look direction is to the south. The fracture is consequently subdued.

An important terrain property that has a strong

influence on the degree of backscatter and hence the resultant signature is the roughness of the surface. Roughness in this context does not refer to macroscale variations such as topographic expression. Instead it applies to 'microrelief' variations which are measured in terms of centimetres and can be caused by differences in grain size (sand, gravel, cobbles) or types of vegetation. For example, the forest on Figure 3.20b, which is composed of tree trunks, branches with different orientations, and leaves is rougher (brighter) than the grassland. Two equations, referred to as the modified Rayleigh criterion, can be used to assess whether a surface can be considered radar rough or radar smooth. A surface is radar rough if:

$$H > \frac{\lambda}{4.4 \sin \gamma} \qquad \text{equation 3.8}$$

and radar smooth if:

$$H < \frac{\lambda}{25 \sin \gamma} \qquad \text{equation 3.9}$$

where H is the surface roughness, λ is the wavelength and γ is the depression angle.

A radar-smooth surface reflects the incident energy away from the radar platform, no signal is returned

Example: A surface has an average roughness of 10 cm. If the surface is imaged by a radar system operating at a wavelength of 23 cm with a depression angle of 45 degrees, would it be classed as a radar-smooth or radar-rough surface?

From equations 3.8 and 3.9 and substituting for λ (23 cm) and sin 45 (0.707), a surface is radar rough if H is greater than 7.4 cm and smooth if H is less than 1.3 cm. Because the surface roughness is 10 cm it would appear bright on the radar image and be classed as a radar-rough surface. For this system for roughness values between 1.3 cm and 7.4 cm an intermediate grey tone would be assigned. It can be seen from equations 3.8 and 3.9 that by, investigating the same surface using different wavelengths and depression angles, the signatures obtained may allow its roughness to be determined quite accurately.

Example: Estimate the roughness of a surface which is classed as radar rough on a radar system that has a wavelength of 3 cm and a depression angle of 30 degrees and radar smooth on a system that operates at a wavelength of 23 cm at a 40 degree depression angle.

For a wavelength of 3 cm and a depression angle of 30 degrees, a radar-rough surface has a roughness greater than 1.36 cm whereas for the system that operates at a wavelength of 23 cm and a 40 degree depression angle, a radar-smooth signature is less than 1.43 cm. Therefore the roughness of this surface lies within the 1.36–1.43 cm range.

and a dark 'radar-smooth' signature is obtained (1, Figure 3.25a). A surface with an intermediate roughness (2, Figure 3.25a) scatters a small amount of energy back towards the sensor (yielding an intermediate tone), whereas a rough surface (3, Figure 3.25a) scatters a large proportion back to the sensor and produces a bright signature.

As has been mentioned previously, a radar-smooth surface will reflect all of the incident radiation away from the radar receiver and will be assigned a black signature. However, if the radar-smooth surface has a favourable orientation with respect to the radar beam, although the incident energy is still reflected, it may be reflected towards the receiving antenna and will be assigned a correspondingly bright signature (Figure 3.25b). This effect may occur for sand dunes that generally have dark signatures except for suitably orientated faces. The same effect may occur for horizontal surfaces which are being imaged at very high (c. 70 degrees) depression angles.

The electrical properties of a material will influence to some extent the radar signature obtained for that material. The extent to which the signature is affected may be determined by the dielectric constant. A material with a low dielectric constant (value of 3) such as dry soil absorbs a large amount of the incident electromagnetic radiation and reflects a small amount. Con-

Example: From the foregoing discussion, what do you think the radar signature for water should be?

You might expect that, because of the high dielectric constant for water, it will have a bright signature. However, an examination of Figure 3.19b shows that the sea is associated with a very low radar return and is radar smooth and thus dark. It is important to realise that the dielectric constant is not the only parameter affecting the signature. It is much less important than surface roughness. A flat stretch of water is radar smooth and even though water has a high dielectric constant, the incident energy will be reflected away from the sensor and will thus appear dark. The bright signature associated with the dendritic drainage patterns shown in Figure 3.20b is not caused by the rivers but by the radar-rough vegetation that is aligned along the banks of the rivers. Thus with ice, snow and water, it is the surface roughness rather than the dielectric properties which allow us to distinguish these features.

Example: What major disadvantage does radar have compared with Landsat MSS 7 images (photographic infrared) in the search for water resources in arid terrain?

In arid terrain, on an MSS 7 image, a water-filled river course will have a black signature that is easily differentiated from a dried-up river course which is often outlined by highly reflective sand. However, on a radar image, water may be black (radar smooth) and a sand-filled river course may also be black. In the 1981 Space Shuttle radar experiment over the Red Sea Hills of Sudan, the wadis (which are devoid of water) are shown black, the same signature as the River Nile. An examination of MSS 7 images of the same area showed that only the Nile had any water flowing along it, all other river systems were barren.

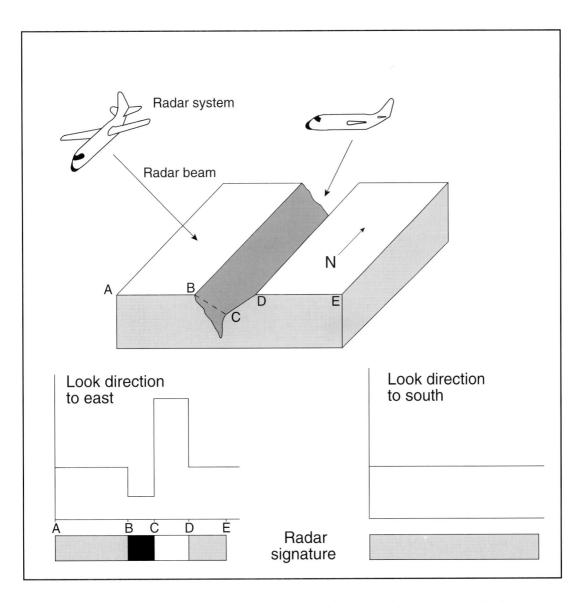

Figure 3.24 Effects of look direction on radar signatures. In general, features trending at a high angle (> 45 degrees) to the look direction are highlighted while features at a lower angle are suppressed.

sequently only a small amount of energy is available to be backscattered to the antenna and a low radar return is obtained. If the soils are moisture laden, because the dielectric constant for water is high (30–80 depending on wavelength), a smaller proportion of the incident energy is absorbed, thus more is available for backscat-ter and a higher radar return can be acquired. Thus, in general, variations in soil moisture should be indicated by variations in signature.

Metallic objects have high dielectric constants and can also be rough at radar wavelengths yielding bright signatures. They also produce high returns

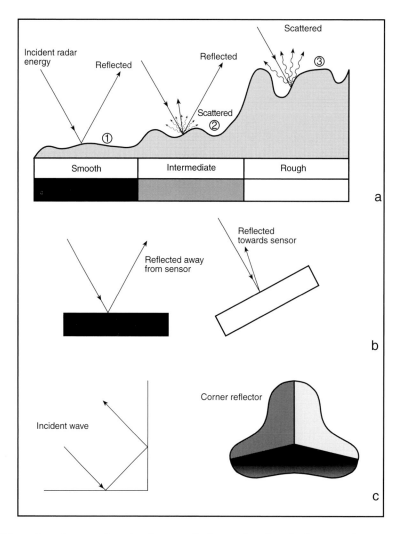

Figure 3.25 (a) Effects of roughness on the radar signature. Radar-rough surfaces backscatter a large proportion of the incident energy and have a bright signature whereas a smooth surface backscatters virtually none and is dark. (b) Effects of orientation of a surface on a radar signature. A radar-smooth surface may, if it is favourably oriented with respect to the incident radar beam, produces a bright radar return. (c) Corner reflectors reflect all the incident energy back towards the radar receiver and consequently are associated with bright signatures.

because they can form corner reflectors (Figure 3.25c). A corner reflector is formed by three surfaces that meet at right angles. An incident radar beam that strikes a corner reflector, irrespective of its surface roughness, will reflect the energy back towards the receiver and produce a bright radar signature.

Buildings, the crossbeams on bridges, and field boundaries may all act as corner reflectors. Small yachts often carry a small corner reflector on their masts in order that larger ships that use radar may detect them. The newly developed Stealth Bomber of the United States Air Force has been designed to minimise radar returns to render it invisible to detection by reflecting radiation away from the receiver and by being coated with special paint that absorbs the radar energy.

Surface Penetration by Radar

Microwave electromagnetic radiation may, under certain conditions, be able to penetrate the surface of sand or soil. The depth of penetration is dependent on the radar skin depth (δ) which is the depth by which the radar pulse has decreased in amplitude to 37 per cent of its initial value. The radar skin depth decreases with increased values of the dielectric con-

stant, which, because the latter is proportional to moisture content, means the drier the conditions the greater the penetration. The radar skin depth also depends on granularity, a deeper penetration being attained for fine sand than for gravel. Penetration is also greater at longer wavelengths and at higher depression angles. In normal conditions, the amount of moisture in the sand or soil limits the penetration to a few centimetres. However, in hyperarid areas, such as in parts of north Africa, penetration to much greater depths (2 to 3 metres) can be achieved.

For a radar-smooth sand-covered area, an incident wave is partly reflected by and partly transmitted through the surface material (Figure 3.26). Where

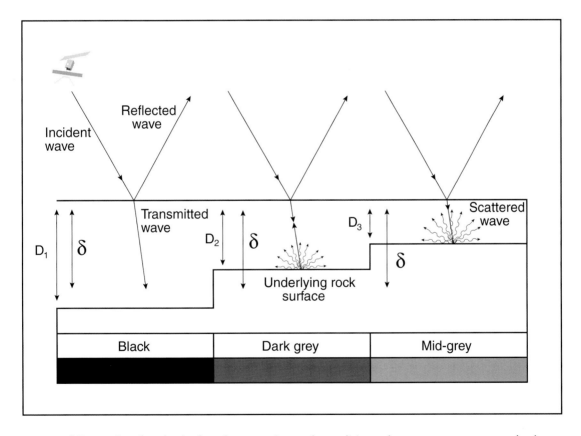

Figure 3.26 Penetration of overburden by radar energy. In very dry conditions radar energy can penetrate overburden to considerable depths. If the depth is too great (D1) then the radar response is smooth (black). However, if the overburden is not too thick (D2 and D3), then scattering from the underlying rock surface may occur and subsurface information can be obtained.

the thickness of the overburden (soil or sand) is much greater than the radar skin depth (D1 > δ), all the transmitted energy is attenuated and the response remains black, i.e. a radar-smooth surface. However, if a rock surface is present at a depth less than the radar skin depth (D_2 or D_3 < δ), scattering may occur from this surface (if it is rough enough) similar to that from an exposed rough surface. These penetrative capabilities have led to the discovery of unknown former river channels and archaeological remains in Egypt and allow the continuity of geological structures beneath the sand to be mapped.

Figure 3.27 (a) Landsat image of igneous intrusions in Mali. (b) SIR-A image of the same area also displaying the intusions but the penetrative capabilities of radar also show the presence of numerous dykes. Width 100 km. Courtesy of NASA/JPL.

Figure 3.27a shows a Landsat MSS 7 (photographic infrared) image of part of Mali in west Africa. The bright areas represent sand with a high reflectance. The most prominent features are circular igneous intrusions. Many of these have a rim that is more resistant to erosion than the centre is. Thus the rock is exposed around the edge but windblown sand fills the centre portion. The same area on radar (Figure 3.27b) appears markedly different. The rim of the intrusions is bright, indicating a radar-rough signature which is due to rock being exposed at the surface. The sand has been penetrated to reveal that an extensive linear array of dykes is associated with the intrusions which go undetected on Landsat. The broad channel has a black signature on radar, thus the sand deposits in this region are thicker than the radar skin depth.

Compression and Layover Effects on a Radar Image

The pulses of electromagnetic radiation transmitted by radar systems have an arcuate wavefront (Figure 3.28). When this beam intersects a steep feature, such as a mountain, the beam strikes the top before it reaches the base of the feature. Consequently the radar return from the top is collected back at the receiver before that of the base. The radar system uses the time delay in the transmission and return of the electromagnetic pulse to measure the distance to the feature. The radar system will produce an image in which the top of the mountain is closer to the receiver than the base and in the resultant image the mountain, which may be symmetrical, will appear to lean

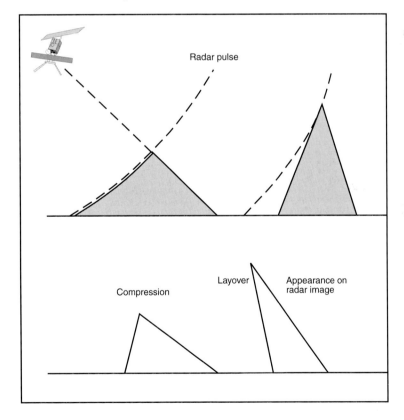

Figure 3.28 Steep symmetrical features may appear asymmetric on a radar image because of the curvature of the radar pulse. This can produce compression for slopes facing the radar transmitter and in extreme cases a layover effect is apparent in which the features 'leans' towards the transmitter.

over towards the near range. This distortion effect is known as layover. It is most noticeable in mountainous terrain but this effect is not constant throughout the image. For a radar system with a large variation in depression angle, the time delays for the returns from the peak and base are greater for a high depression angle than for a lower depression angle, so that the distortion is greater in the near range. If the scene contains terrain with less steep slopes, then compression in the near range may result in symmetrical mountains having a scarp-like appearance (Figure 3.28).

Resolution on Radar Images

The spatial resolution for a radar system is different in the range and azimuth directions. A radar antenna produces short-duration pulses of electromagnetic radiation that intercept various targets in the range direction and also records the return signals scattered from the targets. If two targets are very close together in the range direction, then the returned signals will overlap and the radar system will not differentiate between them. The theoretical slant range resolution (R_s) is dependent on the pulse length (τ) and is given by $(c\tau)/2$ where c is the velocity of electromagnetic radiation. In order to obtain the ground-range resolution (R_g), the slant-range resolution must be divided by $\cos \gamma$ where γ is the depression angle. Thus:

$$R_g = \frac{c\tau}{2 \cos \gamma}$$

equation 3.10

Example: Two buildings are separated in the range direction by a distance along the ground of 35 m. Will these buildings be resolved in the range direction if imaged by a radar system operating with a pulse length of 0.2 microsecond and a 35 degree depression angle?

Substituting from equation 3.10 gives:

$$\frac{3 \times 10^8 \times 0.2 \times 10^{-6}}{2 \times 0.819}$$

$$R_g = 36.6\,\text{m}.$$

The range resolution is greater than the separation thus the buildings will not be resolved.

Equation 3.10 shows that the range resolution for a radar system is better in the far range, i.e. at a smaller depression angle (Figure 3.29). If in the example above the depression angle was 30 degrees then the ground resolution would be 34.6 m and the buildings would be resolved. The resolution may also be improved by shortening the pulse length. However, shorter pulse lengths result in lower energy pulses which may restrict the distance in the range direction that can be imaged.

The ground area illuminated by a radar beam is not of constant width but increases in the range direction, producing a triangular area (Figure 3.29). Thus two objects in the near range (1 and 2) may be resolved whereas two other features which have the

---

Azimuth and Range Resolution Considerations

Both the azimuth resolution and range resolution are dependent on τ, the pulse duration, and γ, the depression angle. However, both resolutions are directly proportional to τ whereas the range resolution is inversely proportional to $\cos \gamma$ but the azimuth resolution is directly proportional to $\cos \gamma$. Consider the situation illustrated below where three

buildings (A, B and C) are being imaged by radar. Buildings A and B are separated in the range direction by 200 m but are not separated in the azimuth direction whereas A and C are separated by 12 m in the azimuth direction but not in the range direction.

```
A □ — 200 m — B □
|
12 m
|
C □
```

The radar system has the following characteristics: a wavelength of 23 cm, antenna length of 2 m, and a pulse length of 1 μs. If the buildings were imaged in the far range with a depression angle of 30 degrees then, from equations 3.10 and 3.11, the range resolution is 173 m and the azimuth resolution is 15 m. If, however, the same buildings were imaged in the near range with a depression angle of 60 degrees, then the range resolution is 300 m and the azimuth resolution is 8.6 m. If the buildings were in the far range then A and B would be resolved but A and C would not whereas in the near range A and B would not be resolved whereas A and C would be. Thus range resolution is better in the far range and azimuth resolution is better in the near range. An ideal radar system would have the capability to image the same area at multiple wavelengths, different polarisation combinations and a range of depression angles.

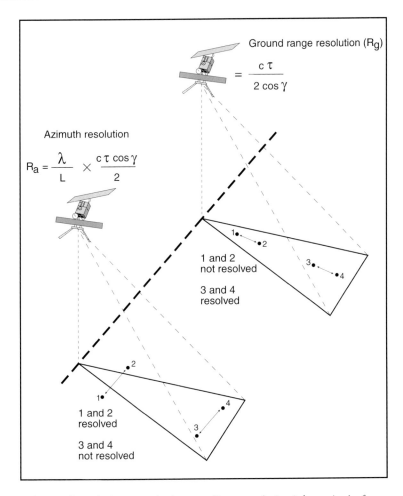

Ground range resolution (R$_g$)

$$= \frac{c\,\tau}{2\cos\gamma}$$

Azimuth resolution

$$R_a = \frac{\lambda}{L} \times \frac{c\,\tau\,\cos\gamma}{2}$$

1 and 2
not resolved

3 and 4
resolved

1 and 2
resolved

3 and 4
not resolved

Figure 3.29 Range and azimuth resolutions on radar imagery. Range resolution is better in the far range whereas azimuth resolution is better in the near range.

same azimuth separation but are in the far range (3 and 4) will not be distinguished. A narrower beamwidth improves the azimuth resolution (R_a). The beamwidth varies directly with wavelength (λ) and indirectly with the length of the antenna (L_a). Lillesand and Kiefer (1994) give the azimuth resolution as (λ/L) multiplied by the ground range which is equal to ($c\tau/2$) multiplied by cos γ. Thus the azimuth resolution is given by the equation:

$$R_a = \frac{\lambda}{L_a} \times \frac{(c\tau \cos \gamma)}{2} \qquad \text{equation 3.11}$$

The azimuth resolution may be improved by either decreasing the wavelength, which may cause problems because shorter wavelength radars are affected by the weather or by increasing the antenna length, though weight and stability considerations restrict the length of the antenna that can be carried by a platform.

Doppler Shift

The Doppler shift is an effect which we have all experienced at some time in our lives. If an ambulance or police car is approaching while using its siren, there is a change in pitch as it passes by. The Doppler shift is the apparent change in frequency of a wave because of the movement of the source of the wave. The frequency of the siren has not changed but only its relative motion. The same principle applies to electromagnetic radiation and is used by astronomers to determine how fast galaxies are moving (Figure 3.30a). For an electromagnetic wave (such as within the microwave where radar systems operate) the frequency (f_o) of an electromagnetic wave moving towards an observer is given by the equation:

$$f_o = \frac{f_s}{(1 - v/c)} \qquad \text{equation 3.12}$$

where f_s is the frequency of the source, c is the velocity of light and v is the velocity of the source towards the observer. As the source moves away from the observer, the observed frequency is given by:

$$f_o = \frac{f_s}{(1 + v/c)} \qquad \text{equation 3.13}$$

where v is the velocity away from the observer. The observed frequency is greater than the source frequency when the source is moving towards the observer and less than the source frequency when the source is moving away from the observer. For an airborne or spaceborne radar system, the forward movement of the platform means in effect that the target apparently moves past the platform in the opposite direction. The SAR begins recording the radar returns when the object enters the beam when, initially, the target is moving towards the platform (1, Figure 3.30b). From equation 3.12, it produces higher frequency returns, has no relative movement at location 2 (frequency returns are the same as the frequency of the radar system) and continues recording as it moves away (3, lower frequency returns). This effectively synthesises a much longer antenna than its actual physical size and correspondingly produces a much better azimuth resolution. Far-range objects are illuminated for a longer time than near-range ones and thus the azimuth resolution is constant for the image.

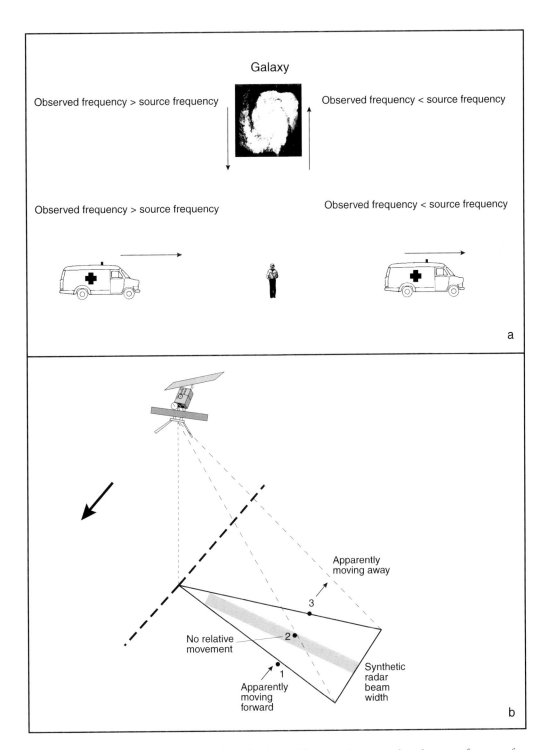

Figure 3.30 (a) Examples of the Doppler shift where the observed frequency is greater than the source frequency for objects moving towards an observer and vice versa. (b) Synthetic aperture radar which employs the Doppler effect to synthesise a narrow radar beam, thus improving the azimuth resolution.

Synthetic Aperture Radar (SAR)

Azimuth resolution can be improved by increasing L_a, the antenna length, though there is a practical limit to the size of antenna that is carried on a platform. However, it is possible, by correctly processing the radar returns coming from the targets, to increase the antenna length *synthetically*. The processing is complex and synthetic aperture radars are inherently more complicated than conventional radars (real aperture radars, RAR). However, the improvement in resolution is so great, synthetically increasing antenna length from metres to kilometres, that most operational radars are synthetic aperture radars. A SAR uses the principles behind the Doppler shift to generate images with high azimuth resolutions.

Spaceborne Radar Systems

The first spaceborne radar images which the public had free access to were obtained by SEASAT. This satellite was launched on 28 June 1978 into a 795 km orbit, (Table 3.6). The onboard SAR operated at a wavelength of 23.5 cm with a 67 degree depression angle for three months before malfunctioning. Although it was primarily designed to obtain data for oceanographic purposes, images were also acquired over land. In addition,

SEASAT also carried two other active microwave systems, an altimeter and a scatterometer. The altimeter transmitted pulses of electromagnetic radiation which, as well as allowing the shape of the sea surface to be calculated to an accuracy of 10 cm, also allowed estimates of the wave height and wind speed to be made. A scatterometer allows the wind speed and direction to be calculated and may also be used to construct backscatter profiles for terrain features at different depression angles.

A series of Shuttle Imaging Radar (SIR) experiments has been performed aboard certain Space Shuttle missions since 1981. SIR-A operated at a single wavelength and a fixed depression angle whereas SIR-B had a variable depression angle (Table 3.6). The most sophisticated of the Shuttle experiments to date has been the SIR-C mission in 1994. This system (which operated in conjunction with an X-band 3 cm radar) could obtain data at two different wavelengths at a range of depression angles with various polarisation combinations. The versatility of this SAR permits the construction of colour radar images which allow a greater amount of information to be extracted for any one area. Plate 3.3 shows a false colour image of Brazilian forest formed by projecting an L band in HV mode in red, a C band in HV mode in green and an X band in VV mode in blue. Pink regions represent rainforest still in its natural state whereas regions

Table 3.6 Characteristics of main spaceborne radar systems

| | SEASAT | ERS-1 | RADARSAT | Shuttle Imaging Radar | | |
				SIR-A	SIR-B	SIR-C
Operational altitude (km)	798	785	800	259	360;257; 224	225
Operational wavelength (cm)	23.5	5.6	5.6	23.5	23.5	5.8; 23.5
Depression angle (°)	67±3	67	variable	37–43	25–75	27–73
Range resolution (m)	25	approx. 26	variable 8.4–73.5	40	16–58	approx.30
Azimuth resolution (m)	25	30	variable	40	20–30	approx.30
Swath width (km)	100	100	variable	50	20–40	15–90
Polarisation	HH	VV	HH	HH	HH	HH; VV; HV; VV

where the forest has been felled and cleared for agricultural purposes are illustrated in shades of green and blue. The bright red areas are due to a heavy rain storm which was detected by the short wavelength band.

RADARSAT, a SAR satellite operated by the Canadian Space Agency, was launched on 4 November 1995 into a near polar 800 km high orbit. The satellite carries a C-band imaging radar (5.6 cm) with a HH polarisation configuration. Although it operates at a single wavelength and polarization (unlike the SIR-C experiment), it is designed to operate continuous for five years using a number of imaging modes which can be varied depending upon the target of interest. Standard format data are obtained for a 100 km wide swath with a depression angle between 41 and 70 degrees with a 25 m (range) and

28 m (azimuth) resolution. A fine-resolution image with an improved range and azimuth resolution of approximately 9 m can be obtained for a narrower swath (45 km) at a depression angle of 42–53 degrees. Alternatively, a large area (510 km wide) can be imaged (ScanSAR wide mode) at a coarser resolution (100 m) at a 41–70 degree depression angle. Images of different widths can also be produced at resolutions between those obtained using the fine-resolution and ScanSAR (wide) configurations.

The repeat cycle for RADARSAT varies depending on the image swath that is obtained. The Earth is covered every 24 days at a standard 100 km wide swath but the maximum ScanSAR format images most of the Earth (north of 80 degrees South) every six days. The Arctic regions can be imaged daily.

Figure 3.31 RADARSAT image of Kahului airport, Hawaii. © Canadian Space Agency/ Agence Spatiale Canadienne (1996). Data received by the Canada Centre for Remote Sensing. Processed and provided courtesy of RADARSAT International.

RADARSAT is thus ideal for monitoring ice movement and the extent of sea-ice, which, apart from its scientific role, is important for shipping in northern latitudes. RADARSAT may obtain its data at approximately 6 a.m. local time when it is travelling south or at 6 p.m. local time when it is travelling north. The look direction in the morning is to the west while in the evening it is to the east. Kahului airport, Hawaii is shown on Figure 3.31. The smooth runways produce a dark signature similar to that from the ocean that contrasts with the intermediate-toned mottled textured signature for the land. The terminal, hangars and other buildings associated with the airport produce bright radar returns, partly because of the materials they are constructed from and also because they act as corner reflectors.

The European Space Agency launched the European Radar Satellite (ERS-1) on 17 July 1991 and ERS-2 on 21 April 1995. They have a 785 km high near polar sun-synchronous orbit with a repeat cycle of 35 days. ERS-1 carries a number of remote sensing instruments. The Active Microwave Instrument (AMI) consists of a SAR and a wind scatterometer. The SAR is a VV system that operates at 5.6 cm and obtains data for a 100 km wide swath with a depression angle of 67 degrees in image mode. Resolution in the azimuth and range direction is approximately 30 m. In wave mode, small areas (5 km x 5 km) are sampled every 200 km along track, which allows the height and direction of ocean waves to be determined. The wind scatterometer uses three radar beams pointing forward, backward and normal to the direction of movement to analyse a 500 km wide swath of the sea surface. The three separate signals obtained by the antennae allow the speed and direction of the wind to be measured. The directional measurements have an accuracy of 20 degrees. Speeds between 4 m/s and 24 m/s can be measured with an accuracy of 2 m/s or 10 per cent. Wind and Wave modes can be operated in conjunction but the wind scatterometer cannot be operated simultaneously with image mode.

ERS-1 also performs passive remote sensing using the Along Track Scanning Radiometer (ATSR). This consists of an InfraRed Radiometer (IRR) which obtains data centred on four wavebands (1.6 μm, 3.7 μm, 10.8 μm and 12 μm) for a 500 km wide swath at a resolution of 1 km. Sea-surface temperatures can be measured to an accuracy of 0.1 K. In addition, a two-channel microwave sounder obtains data on atmospheric water vapour at 0.82 cm and 1.26 cm. ERS-1 also carries a radar altimeter with an accuracy of 10 cm and Precise Range And Range-Rate Equipment (PRARE) which malfunctioned soon after launch. ERS-2 radar instrumentation is similar to that of ERS-1. However, the ATSR has been improved and measures three additional channels, two in the visible and one in the infrared and also carries the Global Ozone Monitoring Experiment (GOME). This is a spectrometer that obtains spectral data in the 0.24–0.79 μm range in very narrow bands (0.4 nm). The data allow the concentrations of ozone, water vapour, nitrogen dioxide and bromine oxide in the stratosphere and troposphere to be determined.

An ERS image of the coast of Portugal taken on 4 October 1994 is illustrated in Figure 3.32. The land-sea contact is well displayed, as are the rugged mountains and the bright radar-rough signature for the town of Oporto. The dark signature at sea is due to an oil spill. The oil, lying on the surface, suppresses wave action, which consequently yields a darker (smoother) tone. Spaceborne radar systems are ideal for detecting such oil spillages because, in the visible and near infrared, the signatures for water and oil are similar and may not be detected by SPOT or Landsat. Radar may also be operated through cloud or at night. Radar images have been used to map underwater sandbanks even though the microwave energy does not penetrate the water. Variations in water depth above the sandbanks may produce variable wave heights which can yield different radar signatures. KOSMOS-1870, a Russian satellite, carried a SAR which was subsequently adapted for the Almaz-1 satellite, which was launched in March 1991 into a 250–300 km orbit but has since ceased operation. The operating wavelength was 9.6 cm in HH mode and it imaged a 350 km wide swath with a resolution of 10–30 m.

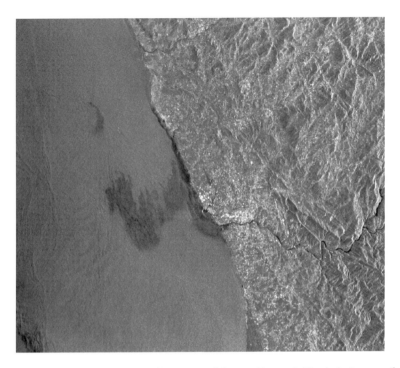

Figure 3.32 ERS image of the coastal region around the town of Oporto, Portugal. The dark signature for the oil spill contrasts well with the unpolluted ocean. Courtesy of Jennifer Fyall © European Space Agency.

Interferometry

SAR images of the same area taken from slightly different positions can be used to create a radar interferogram. The different views can be achieved in a number of ways. The same area can be viewed by the same system at different times from slightly different positions; it can be viewed by different satellites; or the returned radar signals can be recorded by two different antennae separated by a short distance. The latter configuration allows the principles behind interferometry to be easily explained. The signal obtained by one antenna for each individual pixel is out of phase with the signal obtained by the other antenna for the same pixel and this phase difference is related to height variations in the scene. Using these height differences allows a digital elevation model for the region to be constructed. Interferometry patterns may be used in a number of fields of research. Temporal elevation changes in a region can

be determined. Thus the monitoring of snowfields and glaciers can detect whether they are thinning, which might be attributable to global temperature changes. Earthquakes and faults result in the displacement of the land surface; thus the effects of these events may be examined more thoroughly. Small tectonic movements can produce elevation changes which may be the precursors of major volcanic eruptions and earthquakes. The Karakax Valley in China is displayed in Plate 3.4 using data obtained during the SIR-C experiment. Initially an elevation model was produced by interferometry techniques, imaging the same area on different passes, then a colour radar image using L- and C-band data was draped over the model. Such 3-D perspective views allow the relationships between different components of the landscape to be analysed more fully. The range in elevation from the valley to the mountain peaks is 2 km. The Altyn Tagh fault is shown as a lineament running along the right-hand side of the valley at the break in slope just below and parallel to the ridge line.

Other Low-Orbiting Spaceborne Remote Sensing Systems

The National Remote Sensing Agency of India has launched a number of Indian Remote Sensing (IRS) satellites. The first in the series, IRS-1A, had a 904 km near polar sun-synchronous orbit and obtained data in four wavebands (blue, green, red and infrared) with a 73 m resolution using Linear Array Self-Scanning (LISS) imaging systems. The latest in this series, IRS-1C, has a 37 m resolution and also obtains high spatial resolution (approximately 6 m) panchromatic images in the 0.5–0.75 µm range. Images with different spatial resolutions, obtained from Russian space platforms, are now currently available. The KOSMOS satellites obtain photographic data using two camera systems, one of which obtains data for a 265 × 170 km area at approximately 10 m resolution and the other a 2 m resolution for a 165 × 44 km swath. The RESURS satellites have a number of series (for example RESURS-O or RESURS-F) and within each series there are a number of satellite types (for example RESURS-F1, RESURS-F2, RESURS-F3). The RESURS-O carries a pushbroom multichannel scanner which obtains data in three spectral bands (green, red and photographic infrared) at a resolution of 45 m, a radar which operates at 9.2 cm with a 200 m resolution, a multichannel cone-sweeping scanner and a microwave radiometer. More than sixty of the RESURS-F satellites have been launched. They can carry a range of cameras such as the KATE-200, the KFA-1,000, the KFA-3,000 or the MK-4. The KATE-200 is a multispectral system which obtains data in three bands (green, red and infrared) with a 15–30 m resolution for a 270–405 km swath. The KFA-1,000 obtains data with a 10 m resolution in three bands. The MK-4 can obtain data in six bands in the visible and near infrared with a resolution of 6–8 m, though a 2 m resolution has been claimed in panchromatic mode for the KFA-3,000 camera

onboard the RESURS-F3 system. The photographic data obtained by these systems is scanned and digitised in order to convert it to a digital format. Two photographs obtained by Russian satellites are shown in Figure 3.33 and illustrate the detail that can be observed on such images. Individual vehicles can be discerned on the bridge in Bangkok and careful examination shows the presence of ships in the river (Figure 3.33a). The photograph of part of New York shown in Figure 3.33b is dominated by skyscrapers. The tall towers in the left centre of the image form the World Trade Centre.

The Russian polar-orbiting METEOR satellite series fulfils mainly meteorological functions similar to those of the TIROS series of the United States. Instruments carried by the METEOR series have included a thermal infrared high-resolution radiometer, an infrared spectrometer which obtains data in ten channels, an ultraviolet backscattering spectrometer and a Total Ozone Mapping System (TOMS). The Okean satellite series has a number of applications in oceanography. Its instrument complement includes a 3.2 cm wavelength radar which obtains a 450 km wide swath with a range resolution of 1.3 km and a 2.5 km azimuth resolution and low (1 km) and mid-resolution (500 m) multiband scanners. In addition, a microwave (0.8 cm) radiometer obtains data for a 550 km swath with a 15 km resolution. The MIR space station is periodically supplied from Earth with modules which contain equipment for particular experiments. KFA-1,000 cameras aboard the Kristall module allow the acquisition of Earth photographs with a spatial resolution of about 10 m. Other instruments employed on MIR have included radar systems, microwave scanning radiometers, visible and infrared scanners and instruments to measure atmospheric ozone concentrations.

Seven satellites in the Nimbus programme were launched between 1964 and 1972. Nimbus 7, launched into a 955 km near polar sun-synchronous orbit in 1972, carried a number of instruments to measure atmospheric components. One of the most important of these was a Total Ozone Mapping Spectrometer (TOMS). In addition, a Coastal Zone Colour Scanner (CZCS) obtained reflected radiation in five

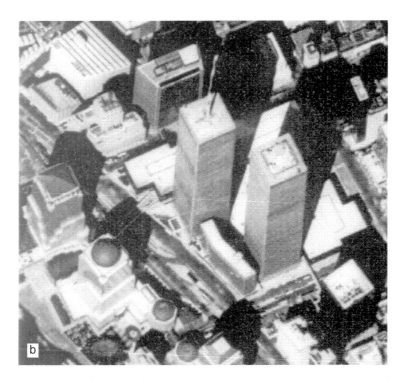

Figure 3.33 Photographs obtained from low-orbiting Russian satellites with an estimated spatial resolution of 2 m: (a) Bangkok; (b) World Trade Center, New York. Courtesy of CEN.

bands in the 0.433– 0.80 μm range and emitted radiation in the 10.5–12.5 μm waveband. This instrument, which allowed phytoplankton concentrations to be measured, obtained data with an 8-bit spectral resolution and a spatial resolution of 825 m for a 1,556 km wide swath. Plate 3.5 shows the phytoplankton concentrations around Tasmania. The lowest concentrations are blue/purple (0.1–0.2 mg/m^3); medium concentrations are green/yellow in colour (0.6–0.8 mg/m^3); and the highest are orange/red (1–10 mg/m^3). The image shows that the phytoplankton are concentrated into distinct zones, often associated with complex swirling patterns. These patterns reflect the movement of water masses with different temperatures, with the highest concentrations in the lower temperature regions. Images such as that shown in Plate 3.5 are invaluable for fishermen because the phytoplankton are at the base of the food chain and fish will tend to congregate where the concentrations are greatest. The SeaWiFS (Sea-viewing Wide Field-of-view Sensor) mission aboard the Orbview-2 satellite, launched in 1997, is also monitoring oceanic phytoplankton variations using eight bands which operate in the visible to photographic infrared range.

The Marine Observation Satellite (MOS-1) was launched by the Japanese National Space Development Agency in 1987. This consisted of a Multispectral Electronic Self-Scanning Radiometer (MESSR) formed of two pushbroom arrays, each of which obtained data in four wavebands (0.51–0.59 μm, 0.61–0.69 μm, 0.72–0.80 μm and 0.8–1.1 μm) with a resolution of 50 m. MOS-1 also carried a Visible and Thermal Infrared Radiometer (VTIR) which obtained data for a 1,500 km wide swath with a 0.9 km resolution in the visible range and 2.7 km in the thermal infrared and a Microwave Scanning Radiometer which obtained data at 23.8 GHz (32 km resolution) and 31.1 GHz (23 km resolution). The TOPEX/POSEIDON satellite was launched in October 1992 into a 1,330 km high sun-synchronous orbit. The onboard sensors include a microwave radiometer for water vapour measurements at 18 GHz, 21 GHz and 37 GHz. A radar altimeter allows the sea-surface height to be determined to an accuracy of 2.4 cm and

the wind speed can be deduced from the strength of the radar return. The Advanced Earth Observing Satellite (ADEOS) operated by NASDA (National Space Development Agency of Japan) was launched on 17 August 1996 into a 830 km high sun-synchronous orbit. The satellite carries a range of sensors including a Total Ozone Mapping Spectrometer (TOMS), an Ocean Colour and Temperature Scanner (OCTS), a scatterometer and an Advanced Visible and Near Infrared Radiometer (AVNIR). The spatial resolution for ADEOS in multispectral mode is 16 m and in panchromatic mode (0.5–0.69 μm) is 8 m. A steerable mirror (± 40 degrees) allows imagery to be obtained from both sides of the satellite and the production of stereoimagery. However, no data have been obtained since mid-1997 because of a malfunction onboard the satellite. An ADEOS panchromatic image of part of Hiroshima, Japan, is shown in Figure 3.34. The area covered is approximately 8 × 6 km though individual buildings may be detected on larger scale images.

The Tropical Rainfall Mapping Mission (TRMM) is a joint NASA/NASDA venture in which a satellite was introduced to a low-inclination (35 degrees), low-altitude (350 km) orbit in 1997. The satellite carries a range of remote sensing instruments including a visible/infrared scanner, a passive microwave imager and a dual wavelength active radar system.

Apart from the radar experiments discussed earlier, photographs of the Earth have been obtained from the Space Shuttle since it was launched. Hand-held photography is obtained by the astronauts using 70 mm Hasselblad or 90/250 mm Linhof Aero Technika cameras. A large format camera (LFC) system was tested on a single mission in 1984. It obtained high-quality photographs on 23 × 46 cm film in black and white, natural colour and infrared. An overlap between the photographs acquired allowed stereoscopic viewing.

The Shuttle has the capability to launch Earth observation satellites. The Upper Atmosphere Research Satellite (UARS) was launched in September 1991 from the Space Shuttle Discovery and it is now in a 585 km sun-synchronous orbit. The orbit is non-polar and has an inclination of 57 degrees. Instruments onboard UARS allow the measurement

Figure 3.34 ADEOS panchromatic image of Hiroshima, Japan. Area approximately 8 × 6 km. Image courtesy of NASDA.

of atmospheric aerosols, temperature, wind speeds and passive microwave emissions.

3.3 IMAGES OBTAINED FROM GEOSTATIONARY SATELLITES

The remote sensing systems discussed earlier obtain data either from low altitude (aerial surveys) or from relatively low orbit (< 1,000 km). The third level at which remote sensing images can be obtained is from geostationary orbit. A satellite in geostationary orbit

is synchronised with the Earth and has a period similar to that of an Earth day (mean solar day length: 23 hours 56 minutes 4.09 seconds) and an orbit that is within the equatorial plane in the same direction as the Earth rotates. Such a geostationary satellite will appear to be stationary overhead, unlike polar orbiting satellites which appear overhead once in every revolution. Geostationary satellites orbit at a height of approximately 36,000 km. Meteorological data have been obtained from geostationary satellites since the early 1960s though the early satellites were primarily designed for transmission purposes. The

Application Technology Satellites (ATS), whose major function was the provision of meteorological information, were launched in the mid-1960s. They were followed in 1974 and 1975 by the Synchronous Meteorological Satellites (SMS) which led to the Geostationary Operational Environmental Satellite (GOES) programme which is currently in operation. The GOES programme consists of the simultaneous operation of five geostationary satellites approximately equally spaced around the equator which together provide nearly total coverage of the globe. The current satellites are GOES West (135 degrees west); GOES East (75 degrees west); Meteosat (0 degrees); Insat (74 degrees east) and GMS (140 degrees east). Malfunctions can reduce the number of operational systems, producing gaps in the coverage though it is possible to redeploy a satellite to a different position in order to compensate for this loss of data. The GOES system obtains data from approximately 80 °N to 80 °S, though the curvature of the Earth produces compression in the high latitudes; thus the best results on the images are obtained between 55 °N and 55 °S. The large segments of the globe imaged by these satellites make them ideal for monitoring wind circulation patterns and global weather systems. As well as sensors monitoring the Earth and its atmosphere, some of the satellites also carry other scientific instruments to measure the Earth's magnetic field and alpha particles and X-rays emitted by the Sun. In addition, distress signals emitted by transponders on the Earth's surface are transmitted and relayed by the GOES system in order to direct rescue missions. The GOES system obtains data approximately every 30 minutes. Such a rapid acquisition is necessary for monitoring weather systems, which are very dynamic and thus change over very short time intervals. The image covered by the satellites is built up by a scanning system called the Visible Infrared Spin-Scan Radiometer (VISSR). The rapid rotation of the satellite causes the sensors to be swept east–west across the globe. Tilting of the satellite allows strips of data to be obtained in order to produce the entire image. Meteosat obtained data in the visible (0.55–0.70 µm) and thermal infrared at 5.7–7.1 µm and 10.5–12.6

Example: Locate the wavebands listed above on Figure 2.5b. What appears odd about the choice of bands?

Remote sensing systems usually have sensors that operate within atmospheric windows, in order that the electromagnetic radiation may reach the surface. The visible 3.9 µm, 11 µm and 12 µm wavebands operated by GOES are within atmospheric windows but radiation with a wavelength of 6.7 µm is almost totally absorbed by water in the atmosphere and does not reach the Earth's surface. However, the purpose of this band is not to make measurements emanating from the surface but to measure the water vapour content of the atmosphere.

µm. However, the latest GOES West to be launched obtains data in five bands, one in the visible and four in the thermal infrared centred at 3.9 µm, 6.7 µm, 11 µm and 12 µm. The visible band is not split into separate red, green and blue components as is the case for SPOT and Landsat, for example, because the component colours are not required when their primary purpose is the detection of clouds and weather systems.

A false colour image of North and South America obtained by GOES-8 is shown in Plate 3.6 in which the visible channel, channel 4 and channel 2 are projected in red, green and blue respectively. The global perspective provided by such images allows the cloud patterns to be investigated. Wind directions and speeds may also be estimated by viewing images acquired at different times. A spiral pattern can be observed west of South America whilst east–west-trending linear formations are seen near the equator. By convention on this type of image, lower temperature regions are represented by higher digital numbers and thus correspond to bright areas on an image. Thus the cold high clouds are highly reflective and appear white, whereas warm low clouds such as can be seen over North and South America are shown in bright red in this band combination. The orange/yellow clouds near the equator

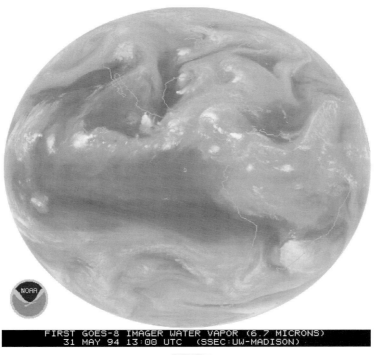

a

b

Figure 3.35 Images from a geostationary satellite showing (a) water vapour and (b) temperature variations for the atmosphere. Images courtesy of Dennis Chesters GSFC/NASA.

are at an intermediate level in the atmosphere. The variation of water vapour with temperature may be evaluated by investigating images such as those in Figure 3.35 which are produced from GOES data. In general, dark regions correlate with warm dry air while brighter areas are colder and wetter.

3.4 OTHER FORMS OF REMOTE SENSING

Although radar is the principal form of active remote sensing, the use of lasers is a well-established technique for providing information in a range of research fields. The technique, in which lasers are used as the illumination source, is referred to as LIDAR (Light Detection And Ranging). Lasers operate at much shorter wavelengths than the microwaves used in radar. Certain substances fluoresce with a characteristic signature when irradiated with a laser operating at a particular wavelength. As the strength of the signature is proportional to the amount of the substance, it is possible to use lasers to measure how much of a particular substance is present. The variations in chlorophyll concentration in the Gulf Stream obtained by NASA's Airborne Oceanographic LIDAR laboratory using a laser operating at a wavelength of 0.532 µm are shown in Figure 3.36a. Concentration is low but constant in the west, very low and constant in the east but extremely variable in the central portion. Thermal data were obtained simultaneously and the concentration is seen to decrease markedly at higher temperatures. Similar laser fluorescence experiments have allowed the thickness of oil films to be measured. This has applications in pollution control because chemical dispersants can be targeted more effectively. LIDAR systems may be employed to obtain very accurate topographic measurements and thus allow elevation changes to be monitored. The Airborne Topographic Mapper (ATM) experiment in Greenland has shown that in the period 1980–1983, most ice accumulation occurred in the western part of the study area while an ablation of 0.75 m was recorded in the east (Figure 3.36b). Other applications for this active remote sensing system have included the measurement of ocean wave profiles, mapping bathymetric variations

and estimating ice thickness. Cosmonauts aboard the MIR space station have also employed a LIDAR, operating at a wavelength of 0.527 µm.

Although satellite data may on occasions be able to provide information on topographic variations beneath the oceans, they are not ideally suited for this purpose. Visible light, which penetrates the farthest, extends only to depths of less than 200 m and thus provides no information about topographic variations or geological features within the deeper parts of the ocean. However, sonar (Sound Navigation And Ranging) techniques allow these features to be mapped. The techniques employed have certain similarities with radar imaging. Pulses of energy are transmitted through the water and, when they intercept a target, are scattered and reflected. These scattered and reflected waves are received by a transducer and the strength of the returned signal allows a black and white image to be constructed. There is one fundamental difference between radar and sonar. Radar uses electromagnetic radiation which travels at 3×10^8 m/s whereas sonar uses sound waves which travel at approximately 1,300 m/s in water. In order to prevent sound waves from the ship's propeller from interfering with the measurements, the sonar device is often towed along behind the ship near the seabed. Narrow strips of data from each side of the sonar device are obtained and the forward movement of the ship allows these strips to be mosaiced into an image. The resolutions obtained vary with the frequency of the sound waves. Shallow-water high-resolution surveys typically obtain a 200 m wide swath with a 10 cm resolution whereas deep-water investigations may image a 30 km wide swath with a resolution of 50 m. Figure 3.37 shows a sonar image of part of the Atlantic sea floor on which a seamount can be observed. (A seamount, in this case the Sumpter seamount, is a volcanic feature which in this instance occurs along the mid-Atlantic Ridge, a spreading geological plate margin where new sea floor is being created.) Illumination is from the top; thus, the seamount signature is bright where a substantial amount of energy has been returned from its flanks, though no information can be obtained from behind it. (This is equivalent to a radar shadow on a radar

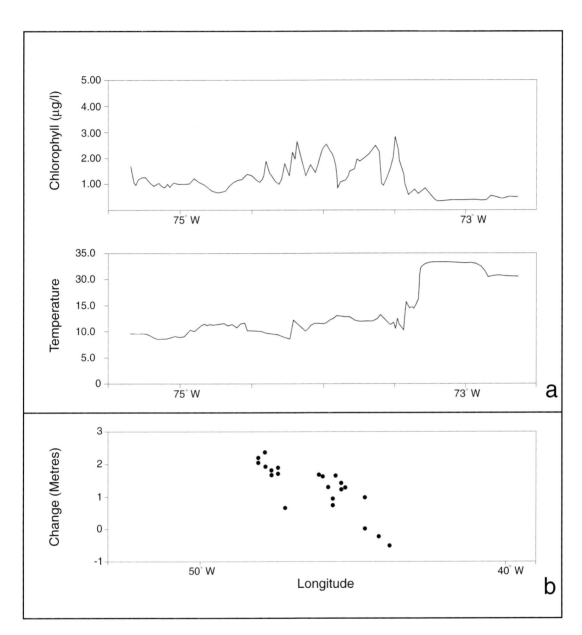

Figure 3.36 Applications of LIDAR: (a) Chlorophyll variations and correlation with temperture. (b) Topographic mapping of changes in the thickness of the Greenland ice sheet. Courtesy of Bill Krabill and NASA.

image.) The seamount, which is 350 m high, tapers upwards from a basal diameter of 2,200 m to 600 m. The sea floor to the west of the seamount is quite rugged while linear features to the north of it represent faults.

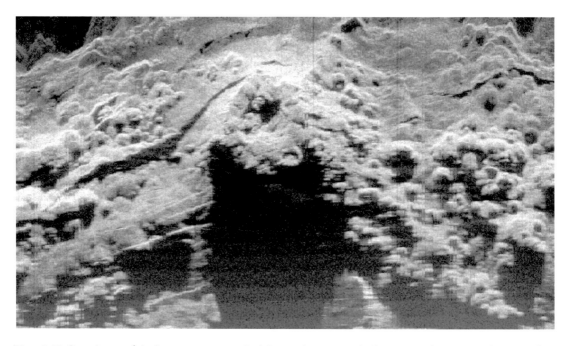

Figure 3.37 Sonar image of the Sumpter seamount. Such images have many similarities to radar imagery. Courtesy of D. Smith, Woods Hole Oceanographic Institute.

3.5 REMOTE SENSING – DEVELOPMENTS IN THE FUTURE

During the last 40 years, the field of remote sensing has expanded enormously and the rate of this expansion is also increasing. Sensor technology has improved greatly, allowing vast amounts of data to be collected. Parallel developments in computing enable these data to be analysed and turned into useful information which provides us with a better insight into our planet. At the time of writing (1998) a number of remote sensing satellites are scheduled for launch.

One of the most important sources of data for the Mission To Planet Earth (MTPE) programme is the Earth Observing System (EOS), a major component of which is a series of satellites scheduled to be launched over the next few years. The satellites are designed to provide information about specific aspects of the planet. For example, AM-1 will monitor pollution in the troposphere and the Earth's radiation balance while COLOR-1 will map oceanic concentrations of chlorophyll and phytoplankton. Aerosols in the atmosphere will be measured by the AERO-1 satellite, and information on atmospheric water vapour content and temperature profiles will be provided by PM-1. ALT-1 will obtain accurate topographic measurements of the sea-surface and ice-cap systems.

Envisat, a near polar sun-synchronous satellite to be operated by the European Space Agency, is to carry a SAR, a radar altimeter, an along-track scanning radiometer and instruments to measure atmospheric temperatures and concentrations of atmospheric gases.

It is to be expected that many of the current space-borne satellite programmes such as Landsat and SPOT will continue, though, as they are government funded, their continuation (or not) is often a function of economic and political circumstances. No launch dates have been included in this section because a quick scan through remote sensing literature of a few years ago shows that projected launch dates are either often not realised or postponed. Technological difficulties in the testing of systems and monetary constraints often control the timing of launches and the instruments that are carried. In addition, disasters like the destruction of the Challenger Space Shuttle or the loss of Landsat 6 in 1993 can completely change any planned scheduling.

The Land Remote Sensing Policy Act (1992) in the United States has enabled private companies to obtain licences to launch and manage their own satellite systems. Such a move is to be welcomed as it prevents governments monopolising and on occasion restricting access to data. However, it is unlikely that such companies will provide a source of *inexpensive* remote sensing data to the wider community. A change in the types of satellites is taking place in conjunction with the opening up of remote sensing to the private sector. Landsat 6 cost about 300 million US dollars, and provided no data. Although governments may sustain such losses, the private sector may be reluctant to provide capital funding on such a scale. However, 'minisatellites' currently under construction and weighing less than 500 kg will be capable of providing better spatial resolution than many of the environmental satellites in operation at present. The cost of such minisatellites will be substantially less than that of the current generation of satellites. New high-resolution sensors will enable new applications of remote sensing to be developed for monitoring urban and social patterns of human societies. It will also promote increased integration into other spatial datasets through Geographical Information Systems technology.

The future of remote sensing is assured. Systems will continue to be built which greatly improve our understanding of the planet. Remote sensing allows us to monitor changes in the climate, map the extent of pollution and the effects of natural disasters, undertake strategic planning for sustained development and much more. It can also be used to indicate where to find the largest shoals of fish in the oceans, which we can then overexploit, or regions of virgin tropical rainforest, which we can then destroy. What we use remote sensing for is ultimately up to us. Some uses to which remote sensing can be put are outlined in Chapter 4.

3.6 CHAPTER SUMMARY

- Aerial photographs can be obtained in an oblique or vertical format. Overlapping vertical aerial photographs may be viewed stereoscopically to produce a three-dimensional effect. They contain radial and scale distortions.
- Bodies which have similar reflectances in the visible may be distinguished on a thermal image if they possess different thermal properties. Important thermal properties include the thermal inertia, thermal capacity and the thermal conductivity.
- Hyperspectral imaging involves the acquisition of many (> 200) bands with very high spectral resolutions.
- Landsat is a series of Earth observation satellites which has been imaging the Earth since 1972. The satellites use a transverse scanner to obtain their data. The multispectral scanner (MSS) obtains data in four bands for the visible and near infrared with a spatial resolution of approximately 80 m. The Thematic Mapper (TM) onboard Landsats 4 and 5 obtains seven bands of data, six with a resolution of 30 m and a thermal band with a 120 m resolution. Landsat 7 can obtain data with a high spatial resolution in panchromatic mode.
- SPOT operates a pushbroom imaging system which can obtain single-band panchromatic images with a spatial resolution of 10 m and multispectral images (three or four) with a 20 m resolution. Whereas Landsat obtains data from the groundtrack beneath the satellite, SPOT has steerable optics which allow data to be obtained from either side of the groundtrack. The same scene may be imaged from different positions and stereoscopic images can be produced.
- The NOAA TIROS-N series are polar-orbiting satellites whose main function is to provide meteorological data.
- Geostationary satellites such as Meteosat orbit within the equatorial plane and provide images at 30-minute intervals for virtually an entire hemisphere. Their primary function is providing data on weather systems.

- Remote sensing data may be obtained in the microwave part of the electromagnetic spectrum either by measuring weak microwaves emanating from the surface or by producing an artificial source of microwave energy.
- Radar systems operate independently of solar illumination, may operate at night and have the capability to penetrate clouds. Major factors controlling the radar signature are the roughness of the surface, slope of the target and the depression angle.
- In hyperarid conditions, penetration of superficial layers by radar may occur and the underlying rock surface may be imaged. The resolution of a radar image may be different in the range and azimuth directions.
- A synthetic aperture radar uses the Doppler shift to produce synthetically a very long antenna which greatly improves the azimuth resolution. Interferometry allows the production of detailed digital elevation models and the detection of small changes in elevation.
- Dedicated satellites for monitoring different aspects of the Earth's environment are scheduled for launch over the next number of years, thus ensuring a continuing source of remote sensing data.

SELF-ASSESSMENT TEST

1 A camera system with a focal length of 112.5 mm is flown at a height of 2,500 m above sea level over rugged terrain with an elevation of 800 m and 200 m. At the higher elevation, two television masts are separated by 23 mm on an aerial photograph and at the lower elevation two buildings are also separated by 23 mm on the aerial photograph. What is the true distance between the two buildings and true distance between the television masts?

2 How are the orbits of Landsat 1 and Landsat 5 similar and how do they differ?

3 What advantages does SPOT have over Landsat?

4 Complete the following table using your knowledge of roughness criteria for radar systems.

Wavelength (cm)	Depression angle (degrees)	Radar rough (cm)	Radar smooth (cm)
3	45		
6.6		3	
23			1.062

5 What is the difference between the depression angle and look angle of a radar system?

6 Why do GOES satellites have sensors that obtain data at wavelengths which are not within an atmospheric window?

FURTHER READING

Arino, O. and Melinotte, J. M. (1995) 'Fire Index Atlas', *Earth Observation Quarterly* 50: 11–16.

Avery, T. E. and Berlin, G. L. (1992) *Fundamentals of Remote Sensing and Airphoto Interpretation*, 5th edition, Englewood Cliffs, NJ: Prentice-Hall. (Chapters 5, 6 and 7)

Campbell, J. B. (1996) *Introduction to Remote Sensing*, 2nd edition, London: Taylor and Francis. (Chapters 6, 7 and 8)

Drury, S. A. (1998) *Images of the Earth: a guide to remote sensing*, Oxford: Oxford Science Publications. (Chapter 2)

Kilford, W. K. (1979) *Elementary Air Survey*, London: Pitman Publishing Limited. (Chapter 4)

Lillesand, T. M. and Kiefer, R. W. (1994) *Remote Sensing and Image Interpretation*, New York: John Wiley and Sons. (Chapters 5, 6 and 8)

Open Universiteit (1989) *Remote Sensing*, Heerlem: Open Universiteit.

Sabins, F. F. (1997) *Remote Sensing: principles and interpretation*, New York: W. H. Freeman and Company. (Chapters 3, 4, 5 and 6)

4

APPLICATIONS OF REMOTE SENSING

4.1 INTRODUCTION

This chapter provides an overview of the environmental disciplines to which remote sensing can be applied. Its aim is to introduce the reader to the diversity of environmental applications, but it is by no means exhaustive in its coverage. The chapter has been designed so that the reader can gain a good appreciation of the subject without being inundated with technical information. The reader is directed to Chapters 2 and 3 for more technical and operational information on remote sensing and the sensor systems. The level of detail or spatial resolution required to investigate particular surface features often influences the way in which a satellite or aerial image is used. The applications of remote sensing will therefore be introduced according to the spatial resolutions of the systems. The chapter is divided into four main sections:

- Applications of low-resolution satellite images
- Applications of medium-resolution satellite images
- Applications of radar satellite images
- Applications using high-resolution images.

More information on detailed case studies of remote sensing applications and technical details of image processing enhancements can be found in the companion volume, *Introductory Remote Sensing: Digital Image Processing and Applications.*

Development of Applications

Initially, satellite sensors were designed with one or a few interrelated applications in mind and therefore were supported on single-sensor platforms, for example the meteorological satellites, the first generation of Landsat and experimental systems such as Seasat and the Coastal Zone Colour Scanner. However, with the development of Landsat Thematic Mapper (TM) and the SPOT series, satellite sensors

Table 4.1 Approximate spectral wavelength regions in which the most commonly used satellites operate, and their environmental applications

Satellite/sensor	Visible		Near infrared (IR)				Thermal IR						Spa Res
Wavelength (μm)	0.45	0.69	0.9	1.1	1.5	1.75	3.5	3.9	5.7	7.1	10.4	12.5	
GMS/Meteosat													2 km
NOAA AVHRR													1 km
Landsat MSS													80 m
Landsat TM													30 m
SPOT 1, 2, 3 XS													20 m
SPOT 4, XS													20 m
Applications	W	W	Pv	Psp	B	V	V	V	V		M	C	
	M	V	V	C		Mo	R	S		R		M	
	S	LU	LU			Mo	Mo		Mo			H	
	M												

V: vegetation; Psp: plant species; Pv: plant vigour; B: biomass; R: rock types; M: meteorology; W: water; LU: land use; Mo: moisture; C: clouds; S: soils; H: heat/temperature; Spa Res: spatial resolution

and the images they produce have been designed to cater for several different applications (see Table 4.1).

More recently, platform design has concentrated on multisensor payloads on a single platform. For example, the European ERS-1, ERS-2 and ENVISAT have application-specific sensors whose products are complementary to those of the other sensors on board. Thus the sensors on board ERS-1 and 2 are designed to collect data for oceanographic and atmospheric phenomena. ENVISAT is designed to continue this mission but to include vegetation and land resources monitoring sensors too, in order to be a true multi-application platform.

However, beyond the turn of the century, there are growing financial pressures and commercial impetus to develop small single-sensor application-specific satellites again, the so-called SmallSats, but with superior spatial, spectral and temporal resolutions to those of their predecessors. This will stimulate a new era in applications development.

4.2 LOW SPATIAL RESOLUTION APPLICATIONS

General Characteristics of Meteorological Satellite Data

Low-resolution satellite data, such as those obtained from Meteosat or NOAA AVHRR, have spatial resolutions of over 1 km. Initially the sensors collecting these types of data were designed to meet spatial-resolution and imaging-frequency requirements for weather forecasting on meteorological satellites. However, since the improvement of the spectral and radiometric resolutions of these data in the early 1980s, meteorological satellite data have been used for numerous terrain analysis and environmental applications. NOAA AVHRR, for example, combines the high radiometric and spectral resolution required for terrain feature discrimination with greater imaging frequency than the Earth resources medium spatial resolution satellites, such as Landsat. The data are relatively cheap (tens

of pounds) and lower spatial resolution reduces data quantity and thus data-processing time. Applications include:

Meteorological

- Cloud cover and weather forecasting
- Atmospheric monitoring and climate change
- Drought and rainfall monitoring
- Storm event analysis
- Monitoring oceans and water bodies.

Environmental

- Vegetation assessment
- Natural hazards monitoring
- Geology.

Meteorological Applications of Low Spatial Resolution Images

Cloud Cover and Weather Forecasting

The key to meteorological uses of low-resolution satellite image data is to determine the amount of cloud, its pattern and type. This is known as neph-analysis. Cloud height can also be determined from a knowledge of cloud types and temperature measurements from thermal infrared satellite images. Ascertaining cloud temperature and combining this with a knowledge of typical seasonal atmospheric conditions of a study region can enable analysts to determine whether a cloud is likely to be producing rainfall or not. This satellite-derived information can then be combined with ground-based radar and conventional weather-station records to enable cloud base to be determined and rainfall amount and spatial extent to be estimated.

From nephanalysis, other meteorological conditions and weather systems can be identified and wind direction and speed can be estimated. Figure 4.1 illustrates a sequence of Meteosat images in February 1994 showing the development of a depression over the Atlantic approaches to the UK which brought sleet and snow showers as it moved over southern

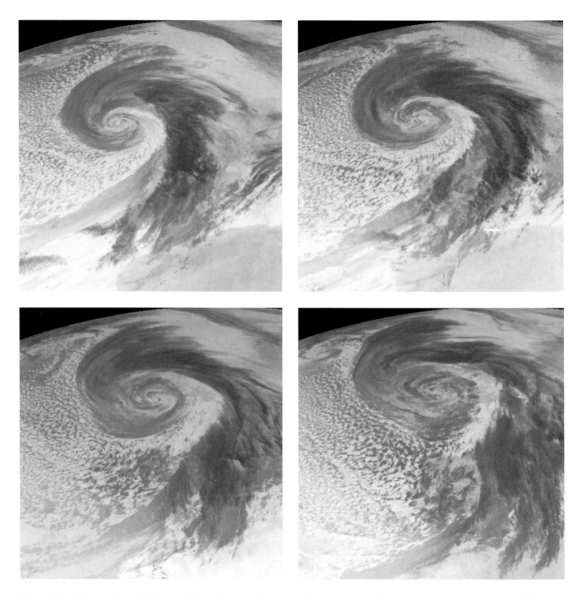

Figure 4.1 A time series of Meteosat infrared images showing a depression developing over Britain on 2 and 3 February 1994. Courtesy of Remote Sensing Unit, University of Bristol and EUMETSAT.

Britain. Continental-scale images can be provided by geostationary satellites (Figure 4.2a and 4.2b). The key cloud phenomena in weather forecasting – fog, convective cloud storms (such as thunderstorms), frontal systems and cyclonic systems – can all be identified by nephanalysis of Meteosat and NOAA AVHRR images. Examples of NOAA images are illustrated in Figure 4.3a, which shows an anticyclone over France, and Figure 4.3b, which displays high-level alto-cumulus and altostratus cloud over Britain on 29 October 1996.

Since 1986 the UK Meteorological Office has

Figure 4.2 Meteosat image bands of Africa and the eastern Atlantic: (a) visible, (b) infrared.

been using the FRONTIERS precipitation 'now-casting' computer system and its descendant, Nimrod (Meteorological Office 1997), to assist

in producing its daily weather forecasts. Every half an hour visible and thermal infrared images are used from both Meteosat and NOAA AVHRR to

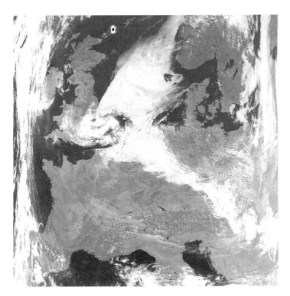

Figure 4.3 Typical AVHRR imagery from NOAA 14: (a) shows an anticyclone over France 24 September 1997 and (b) shows high-level alto-cumulus and altostratus cloud over Britain 29 October 1996. Courtesy of Natural Resources Institute, University of Greenwich, and EUMETSAT.

identify weather systems, supplement and clarify measurements made from ground radar and rain gauges to produce the weather bulletin and forecast (Plate 4.1).

Atmospheric Monitoring and Global Climate Change

Meteorological satellite data archives have existed for nearly 40 years. The minimum period of observations and measurement deemed acceptable

The Ozone Hole

Ozone, a molecule formed of three oxygen atoms (O_3), is a small component of the atmosphere (see Table 3.1). Although it is a relatively minor constituent, its presence is vital because it absorbs incoming ultraviolet radiation, which is a factor contributing to skin cancer in humans. Small

reductions in stratospheric ozone levels may lead to a 2–5 per cent rise in the number of skin cancers. Ultraviolet radiation can also penetrate the top few metres of sea-water where phytoplankton congregate. The phytoplankton are at the bottom of the oceanic food chain and a change in their productivity could affect an entire ecosystem. Ozone can be

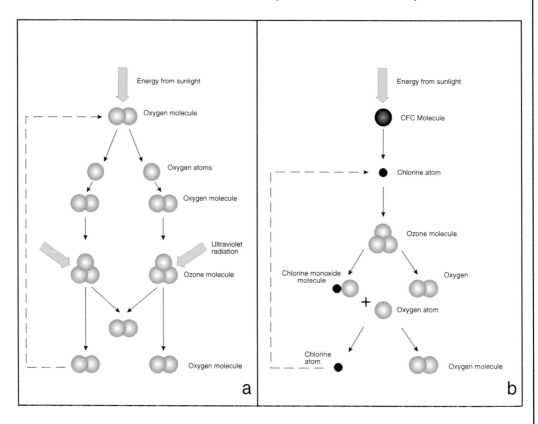

Figure 4.4 (a) Natural balance between the breakdown of oxygen molecules and the formation and dissociation of ozone. (b) Breakdown of ozone resulting from introduction of CFC molecules.

produced by splitting an ordinary oxygen molecule (O_2) into two separate atoms which in turn collide with other oxygen molecules to form ozone molecules, which are concentrated in the lower stratosphere at heights of between 19 km and 23 km. Absorption of ultraviolet radiation by an ozone molecule causes it to split, releasing a single oxygen atom which collides with another oxygen atom to form a molecule of oxygen (Figure 4.4a). This process thus produces more oxygen molecules, which can return through the cycle and will in turn be converted to ozone. Over time a natural balance between ozone and oxygen has been established. However, this natural balance has been disrupted in recent decades by the introduction of chlorine- and fluorine-based chemicals such as chlorofluorocarbons (CFCs). The CFC molecules migrate to the stratosphere where they too are broken down by incoming radiation. A chlorine atom released from a CFC molecule will react with ozone to form an oxygen molecule and a chlorine monoxide molecule. The chlorine monoxide molecule can then capture a free oxygen atom (which previously would have combined with an oxygen molecule to form ozone) which then breaks down to an oxygen molecule and releases the chlorine atom to destroy another ozone molecule (Figure 4.4b).

The introduction of the Total Ozone monitoring experiment on the Nimbus 7 satellite in the 1970s, which had the capability of measuring ozone concentrations, has shown the existence of an 'ozone hole' centred over Antarctica which is at its most extreme in early spring when the photochemical reactions become more active after the dark southern winter. A similar depletion in ozone has now been recognised for the northern hemisphere.

for climatological pattern analysis is 25 to 30 years. Thus, in the last 10 to 15 years, climatologists have become interested in studying satellite data to look for evidence of climate fluctuations and change. The variations in the concentration of atmospheric ozone are currently of major concern.

Data from the Total Ozone Monitor are shown in Plate 4.2. The concentration of ozone over Antarctica in 1979 is shown in Plate 4.2a, where high values are shown in red and low values in blue. The image shows that ozone over Antarctica was relatively high in 1979. While providing important information, this single image does not tell us how the concentration is changing. The image taken in 1992 (Plate 4.2b) shows that the concentration of ozone has decreased greatly over Antarctica. These images first alerted us to the ozone depletion in the atmosphere. A new environmental sensor on board ERS-2 called GOME (Global Ozone Monitoring Experiment) was the first European space sensor for monitoring ozone on a global scale.

Drought and Rainfall Monitoring

ARTEMIS (African Real Time Environmental Monitoring Information System) is a system developed by the UN FAO (United Nations Food and Agriculture Organisation) which routinely produces maps of rainfall estimates for 2.4 km squares covering the African continent using Meteosat data (Plate 4.3). In the lowest layer of the atmosphere (the troposphere), atmospheric temperature usually decreases with height above the Earth's surface. The top layers of a convective cloud may therefore be several tens of degrees Kelvin cooler than the base layer. If the water vapour held in the top layers of the cloud is cooled enough, it will condense to form rain droplets and thus rainfall. This change in temperature with height in the atmosphere is used in nephanalysis of thermal infrared images to determine whether a cloud is producing rain or not.

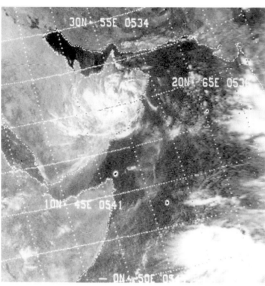

Figure 4.5 Detection of tropical storms and cyclones on low-resolution satellite imagery. This is a tropical cyclone that formed over the Arabian Sea between 12 and 14 June 1977. Courtesy of Remote Sensing Unit, University of Bristol, original data from NOAA.

Table 4.2 Comparison of number of tropical cyclones and severe storms detected over the Arabian Sea by meteorological satellite with conventional and ship recordings

Satellite record				Conventional record		
Year	Month	Dates	Duration (days)	Month	Dates	Duration (days)
1967	May	10–15	6			
	June	12–14	3			
		18–22	5			
	Sep	22–29	8			
1968	May	25–27	3			
	Aug	05–07	3			
1969	May	19–22	4			
	May/June	30–07	9			
	June	25–30	6			
	July	20–22	3			
	Aug	16–19	4			
		22–26	5			
	Oct	20–23	4			
	Nov	17–19	3			
		27–29	3			

Satellite record   *Conventional record*

Year	Month	Dates	Duration (days)	Month	Dates	Duration (days)
	Dec	03–09	7			
1970	May/June	28–01	5	May/June	31–04	5
	Aug	07–10	4			
		27–29	3			
	Sep	05–10	6			
	Nov	21–25	5	Nov	19–29	11
1971				Sep	24–28	5
	Oct	27–31	5	Oct/Nov	27–01	6
	Dec	14–19	6			
1972	June/July	29–02	4			
	Oct	22–24	3	Oct	24	1
1973	June	05–10	6			
	Dec	26–28	3			
1974	April	13–17	5	April	12–17	6
	May	17–22	6			
	Sep	19–24	6			
1975	May	01–11	12			
	June	24–26	3	Oct	22	1
1976	NONE	SIGHTED		NO	RECORDS	
1977	June	08–14	7	June	09–13	5
	June/July	29–02	4			
	Aug	08–10	3			
	Sep	03–06	4			
	Oct	19–22	4	Oct	20	1
	Nov	11–21	11	Nov	12–23	12
1978	IMAGE	DATA	MISSING	Nov	05–13	9
				Nov	24–29	6
1979				May	12	1
	June	16–22	7	June	16–20	5
	Aug	09–14	6			
				Sep	18–24	7
				Nov	13–17	5
1980	June	20,24–26	4			
	Oct	12–18	7			
	Nov	16–18	3			
1981	Oct/Nov	28–03	7	Oct/Nov	28–03	7
1982	Feb	20–22	3			
	June	17–20	4			
	Sep/Oct	30–02	3			
	Nov	06–09	4	Nov	05–09	5
1983	Aug	08–12	5	Aug	10	1
1984				May	26–28	3
	Dec	03–07	5	Nov/Dec	28–08	11
1985	NONE	SIGHTED		NO	RECORDS	
1986	June	05–12	8			
1986	Oct/Nov	31–03	4			
	Nov	07–11	5			
1987	June	06–12	7	June	05–09	5
	Dec	08–11	4			

The full NOAA 35 mm microfilm archive of meteorological satellite data available from 1967 to 1987 has been used with conventional and ship recordings over the same time period.

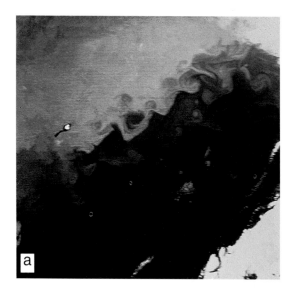

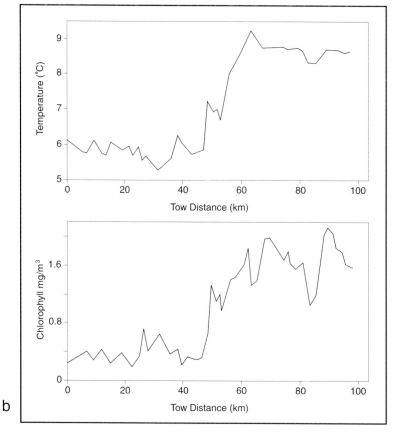

Figure 4.6 The Arctic Front in the North Atlantic where the cold water from the Arctic Ocean meets the relatively warmer water from the Atlantic (NOAA AVHRR August 1986). Courtesy of NRSC, original data: NERC Computer Services, Plymouth University. (b) Thermal and chlorophyll sea data graphs collected by ship measurements. Courtesy of NRSC; original data Plymouth Marine Laboratory.

Storm Event Analysis

Tropical cyclones form over remote ocean areas, but can last several days and therefore travel considerable distances during their life cycle, often reaching land before dissipating. Formation of such cyclones can be seen on low spatial resolution satellite images, but are not always detected by conventional observations (Figure 4.5). A series of images such as these can be analysed by hand or computer to produce tracks of hurricanes and areas of devastation. Table 4.2 shows the results of a study of this kind, listing the cyclones evidenced on meteorological satellite data of the Arabian Sea against conventional ship or aircraft records over the same time period. Considerably more storms were identified by satellite data than by the conventional record.

Global Ocean and Water-Body Monitoring

Processing the data obtained by the thermal infrared channels of NOAA AVHRR allows the production of Sea Surface Temperature (SST) maps, which have been used to locate thermal pollution, warm ocean currents and thus confirm global ocean circulation patterns. Shoals of fish and sites of algal blooms have also been located by the use of sea-surface temperatures derived from NOAA AVHRR.

Figure 4.6a shows the Arctic Front in the North Atlantic where the cold water from the Arctic Ocean meets the relatively warmer water from the Atlantic. The NOAA AVHRR channel 4 image (obtained in August 1986) shows a steep temperature gradient from cold water (which is pale grey) mixing with meso-scale eddies of the warmer water which are darker in colour. Thermal infrared and chlorophyll data graphs show sea data collected by ship measurements (Figure 4.6b).

Satellite monitoring of sea-surface temperature over the last 10 years has enabled ocean and global climate modellers to identify the reverse in the El Niño ocean current in the southern Pacific. This enabled short-term, yet unheeded, warnings of drought in western Pacific and Southeast Asian countries that led to the devastating forest fires of 1997 and 1998 and to the severe snow storms and floods experienced in South American countries. Since 1997 SeaWiFS has been yielding valuable information on ocean colour.

Environmental Monitoring Applications of Low Spatial Resolution Images

Vegetation Assessment

The vegetation index, which is a measurement of vegetation greenness or health, has been developed as a method of enhancing the visibility of healthy vegetation on satellite images and is now a key tool for environmental monitoring. Vegetation indices are formed by ratioing and combining the red and near infrared images of a sensor dataset and are used for vegetation health, primary production, broad-scale vegetation assessment and vegetation mapping. They can be applied to enhance the vegetation, both in natural rangelands and agricultural landscapes. NOAA AVHRR datasets are used for global environmental monitoring applications which require a general view over a large area, from which 'trouble spots' can be targeted, either for Landsat TM analysis or for detailed ground surveying. Thus the use of such images allows scarce resources to be effectively targeted. The low-resolution vegetation index, together with meteorological images analysed for rainfall, can give warning of impending drought or crop failure and show the regional extent of these disasters better than any ground survey or high-resolution satellite study. The global vegetation distribution maps seen in modern atlases are often made from mosaics of low-resolution satellite-derived Normalised Difference Vegetation Index (NDVI) images.

Plate 4.4 shows the vegetation dynamics over a year for the continent of Africa. The images have been produced from the NOAA Global Area Coverage data (GAC) Normalised Difference Vegetation Index product remapped to the Mercator projection. Plate 4.4a shows the NDVI monthly average product for Africa for the month of July 1985 as a single-band image coloured in green. Plate 4.4b shows the monthly average NDVI scenes for January in red, May in green and September in blue superimposed on one another to form a false colour composite. Applying the theory of mixing the colours of light, the image can be interpreted as follows:

- Areas that are white have vegetation in January, May and September, i.e. all year round.
- Areas that are cyan (or turquoise) have vegetation in May and September but not January.
- Yellow areas have vegetation only in January and May.
- Areas that are magenta (pink) have vegetation in January and September but not May.

Note the blue line across the northern Sahel/southern Sahara region of Africa. This shows that vegetation is only present in this region in September, which is just after the west African rainy season of July and August. This indicates a short-lived seasonal flush vegetation for a month or two after the rainy season before drought hits the area again. Since the droughts and famines in this region in the early 1980s, products such as these have been developed by the world food monitoring organisations to assist in future drought warning systems. Since 1998 the VEGETATION instrument aboard SPOT 4 has provided data for 2250 km wide swaths which can also be turned into NDVI maps.

Natural Hazards Monitoring

Since many natural hazards revolve around meteorological extreme events, meteorological satellite images often provide the first warning of these hazards and possible impending disaster. For example, cloud systems causing heavy and prolonged rain leading to flooding, or long periods of rain-free weather evidenced on cloud-free images and shown by changes in vegetation response on these images can give warning of impending drought or bush fire risk.

Meteosat (visible band) and NOAA AVHRR data have been used to monitor floods in Africa (Legg 1989). Meteosat can give a near real-time monitoring capability of floods but its spatial resolution is poor whereas AVHRR provides more detail but less frequent updates on the situation. Organisations such as ESA ESRIN, Italy, and the Natural Resources Insti-

tute of the University of Greenwich, UK, use the thermal channels of NOAA AVHRR to detect bush fires in tropical forests and semi-arid rangelands as part of their overseas aid environmental monitoring programmes. For example, NOAA-12 detected 24,000 fires in the Brazilian rainforest during 1997 and 1998, and satellite monitoring gave the full extent of the uncontrolled Indonesian forest burning in 1997 and 1998. In both cases the satellite information enabled targeting of fire service resources and an assessment of the extent of the devastation. However, it is impossible to monitor bush fires in cloudy weather with optical images.

Geological Applications of Low-Resolution Images

Single images from Earth resources satellites such as Landsat or SPOT can be used to map areas to scales between 1:50,000 and 1:250,000. These images can also be mosaiced together to examine larger areas at scales between 1:500,000 and 1:2,000,000, but the cost of images and the complex mosaicing process limits such surveys to small regions. Meteorological satellite data can be used to provide more generalised views of much larger areas, often avoiding the need for mosaicing. Geological image maps of whole continents have been produced from images from the AVHRR sensor on NOAA polar-orbiting satellites, although, only major geological features can generally be seen. However, thermal data from the AVHRR have been employed for volcano monitoring. These data are relatively cheap, but mosaic construction required for continental mapping is still an expensive process and the need for cloud-free images can be a problem with images taken in different seasons.

4.3 MEDIUM SPATIAL RESOLUTION SATELLITE APPLICATIONS

General Characteristics of Medium Spatial Resolution Data

Medium-resolution satellite data have spatial resolutions of less than 100 m. These data are available from satellites in low polar orbits, some of which are closer to the Earth's surface than the meteorological satellites. However, image frequency is lower, images being acquired only once or twice a month over the same location on the Earth's surface, rather than several times a day with low-resolution meteorological satellites. The two most commonly used sources of data are obtained from Landsat and SPOT.

Medium (and high) spatial resolution data tend not to be used for weather forecasting as the repeat cycle is too long and the scene coverage is too small for analysis of entire weather systems. Most medium-resolution satellites do not have suitable spectral bands for cloud analysis. However, Landsat TM images, which have thermal infrared and visible bands, have been used for meteorological applications where meteorologists have needed to study severe weather events in detail, for example tropical storms and floods.

Environmental Applications of Medium Spatial Resolution Data

Landsats 1, 2 and 3 carried a multispectral scanner with a spatial resolution of 80 m. At this resolution much more detail of the Earth's surface can be seen than on meteorological satellite data. The data were employed for geological mapping. Observation of Mount Etna, Sicily, in July 1979 using MSS band 7 enabled the construction of a map showing lava flow extents prior to 1900, 1970 and 1971 by workers at the National Remote Sensing Centre in the UK. Landsat MSS has also proved useful for deforestation 'before and after' studies. They show virgin tropical rainforest in the Amazon Basin during the mid-1970s in areas which now show comb-like clearance features penetrating into the forest areas in a systematic manner. These Landsat MSS data are old enough to record the former extent of the virgin rainforest. Landsat MSS can give details of vegetated areas including those belonging to the human agricultural infrastructure, but this is limited to regions where field sizes are fairly large, such as in the US and Canada, and not in northwest Europe. Landsat MSS images were also used quite extensively for ocean navigation in areas

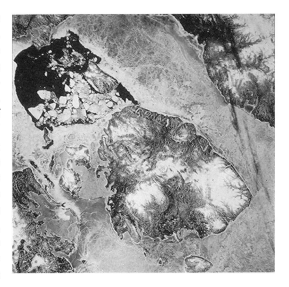

Figure 4.7 Landsat MSS scene showing how satellite imagery can be used for sea navigation in frozen sea areas, Cornwallis Island, Canada, June 1973. Image courtesy of NRSC, original data ESA 1973.

prone to sea-ice before radar satellite data were regularly available (Figure 4.7).

Over the last two decades Landsat Thematic Mapper data have undoubtedly expanded the versatility and range of environmental applications for satellites.

Oceanography and Marine Applications

Sea-surface temperature, algal blooms, estuarine sediment plumes, oil spill and other pollutant contamination plumes have all been successfully detected by Landsat TM when images are available and cloud free at the time of the event. Indeed, it was the thermal pollution in the Chernobyl Lake detected by Landsat TM that first alerted the world to the nuclear power station disaster. However, to monitor the effects of pollution, image acquisition has to be timely and adequate monitoring requires image delivery more often than the Landsat TM or SPOT overpass frequencies. Oceanographical and hydrological studies that use SPOT images are limited because it has neither a thermal infrared band nor a blue visible band.

Vegetation Mapping and Agriculture

The false colour combination of bands 3 (visible red), 4 (near infrared) and 5 (mid infrared) of Landsat TM is frequently used in vegetation/land cover studies. If band 4 is put in the red computer display, 5 in green and 3 in the blue, differences in vegetation health and cover are emphasised. A false colour composite image can be digitally processed in order to produce an image classified into various land uses (Plate 4.5).

It has been possible to assess crop health, type and area by using Landsat TM images since the early to mid-1980s. More recently, agricultural subsidy fraud detection, illegal crop detection and precision farming have added to the portfolio of agricultural applications.

The National Remote Sensing Centre Agricultural Monitoring Project uses cloud-free scenes of Landsat TM and SPOT coverage throughout several growing seasons to ascertain which crops are growing in each field in given areas of the UK. The introduction of an extra infrared band on SPOT 4 has increased its effectiveness in vegetation studies. SPOT has been found to be better than Landsat TM at defining individual field boundaries and small parcels of land of only a few hectares in size, though the area covered by a SPOT scene is only one-ninth of a TM scene. The images are enhanced and interpreted for crop/vegetation cover types. Precision farming is another new application whereby Landsat TM and SPOT images are used to find areas of poor crop development in conjunction with accurate location information from GPS (Global Positioning Systems), so that areas of poor soils can be identified within crop fields. Farmers can then concentrate their fertilising activities on these areas only and not waste fertiliser on areas that already have good soils.

Geology and Soils Mapping

Landsat TM, with its information in the mid-infrared bands, has provided a wealth of information for geological mapping and exploration, often saving companies long and expensive field surveys in areas that are difficult of access. SPOT has fewer bands but a superior spatial resolution, allowing individual faults

Figure 4.8 A prominent NE–SW-trending fault on Landsat TM imagery. Other shorter linear structures in this image also indicate faults or geomorphological features. Image width approximately 15 km.

and rock layers to be mapped. Geology is the most static of the Earth surface features to be observed from satellite, so that a cloud-free image from 1984 is generally just as good for geological mapping as one from 1998, and geologists do not have the problems that other environmental scientists have concerning image-capture frequency. However, in vegetated areas, the nature of the vegetation is often related to the underlying rocks. Thus images obtained at particular times of the year (such as the early part of the growing season), whilst showing vegetation variations, may indirectly provide information on underlying lithologies. Many geological structures, such as faults, are delineated by a topographic signature which is often detectable on a low-angle illumination TM image. Figure 4.8 shows a prominent NE–SW-trending fault. Other shorter linear structures in this image also indicate faults or geomorphological features.

Other Applications

SPOT was commissioned by the military for surveillance purposes during the Gulf War in 1990–1991.

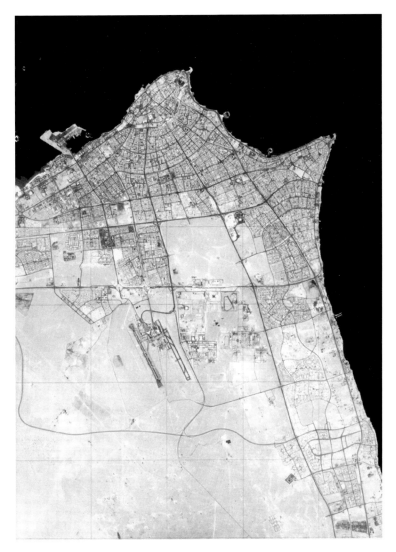

Figure 4.9 An image map of Kuwait derived from SPOT panchromatic imagery. Image courtesy of NRSC, original data ESA 1990.

The military used civilian satellite data from Landsat TM and SPOT to give the first evidence of the oil-well fires during the war and to observe the environmental devastation of Kuwait. In recent conflicts SPOT has been used for arms verification. Its images have also been used by the military for charting previously unmapped areas. Figure 4.9, for example, shows an image map of Kuwait derived from a SPOT panchromatic image.

As spatial resolution has been improved from 80 m with Landsat MSS to 30 m with Landsat TM and the panchromatic 10 m resolution with SPOT, so various researchers have attempted to map urban areas with satellite images. Probably the most successful of these studies are those that look at the urban–rural change in land use during the 1980s and 1990s.

4.4 APPLICATIONS OF RADAR SATELLITE IMAGES

During the late 1970s and early 1980s SEASAT and the Shuttle Imaging Radar proved that radar imaging from space was particularly useful for environmental observation. Since then time and investment have been put into the development of the operational space radar-imaging programmes of ERS and RADARSAT.

The ERS missions have been designed to target several key scientific applications of remote sensing, some of which have been poorly catered for in the past (Table 4.3 and Figure 4.10). Open and polar ocean monitoring and regional seas are a priority. ERS active and passive sensors have been specifically designed to take measurements associated with ocean circulation, elevation (Figure 4.11), tides, ocean-atmosphere interactions and associated energy transfers.

The Along Track Scanning Radiometer has been used to produce thermal cloud maps of tropical cyclones as well as monitoring temperature variations in important sea currents such as El Niño and the Gulf Stream. Coastal processes and shallow water

bathymetry, polar sea and land ice dynamics (Plate 4.6) are also priorities of the ERS missions. However, the ERS missions have contributed valuable data to many other environmental applications. These include extreme meteorological events, contributions to climatological databases, information on crop growth, deforestation (Figure 4.12), hydrological processes, geology and geomorphology, particularly in cloud-covered parts of the world. In addition, ocean-bed topography may be mapped and earth movements associated with earthquakes and volcanoes can be detected.

An important feature of radar satellite data is the ability to differentiate smooth and rough surfaces. This enables the detection of variations in sea-surface tension and roughness caused by oil slicks from tanker spillages. Figure 4.13 shows the oil slick from a Japanese oil tanker on a RADARSAT SAR image. A similar example, but using ERS data, is shown in Figure 3.32.

4.5 APPLICATIONS OF HIGH SPATIAL RESOLUTION IMAGES

Aerial Photography Applications

Although the satellite images illustrated in this book, which display large areas, are visually very impressive,

Table 4.3 Applications for Synthetic Aperture Radar (SAR) images

Application	Parameters provided by SAR imagery and radar instruments
Weather forecasting	Wind speed/wave height data
Sea state forecasting	Wind speed and wave height data
Ship navigation	Wind speed and wave height data
Sea ice monitoring	Image of surface texture
Oil slicks/pollution	Image of surface texture
Coastal processes	Image of surface texture
Land applications	Image of surface texture, height information from interferometry
Land ice	Image of surface texture
Change detection (land)	Multitemporal image
Ocean currents	Image of surface texture
Sea bed topography	Image of surface texture
Wind fields	Image, wind speed and wave height data
Wave fields	Image, wind speed, wave height and altimeter data
Polar ocean monitoring	Image, wind speed, wave height and altimeter data

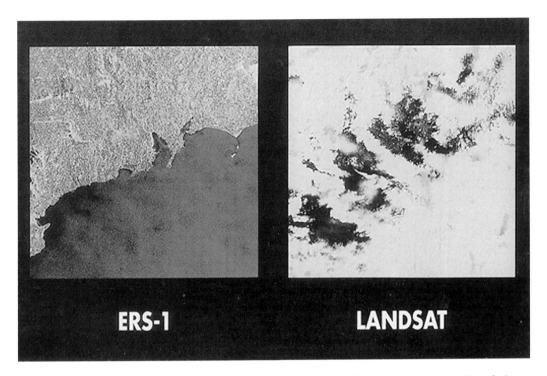

Figure 4.10 A comparison of an ERS-1 SAR image and a Landsat TM image of the same area of coast, Waterford, southern Ireland, on the morning of 9 August 1991. This demonstrates the all-weather capability of the long-wavelength radar satellite sensors. Copyright ESA 1991, 1992, original data distributed by Eurimage.

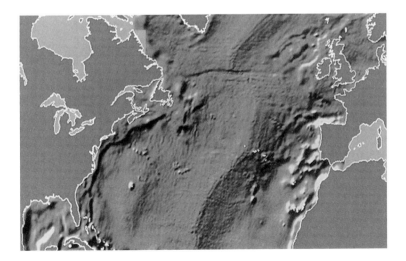

Figure 4.11 Mean sea-surface elevation over the North Atlantic derived from radar altimeter data. Sea-surface topography often gives valuable information about ocean-bed topographic features, ocean currents and the Earth's gravity field. Copyright ESA 1991, 1992, original data by Eurimage.

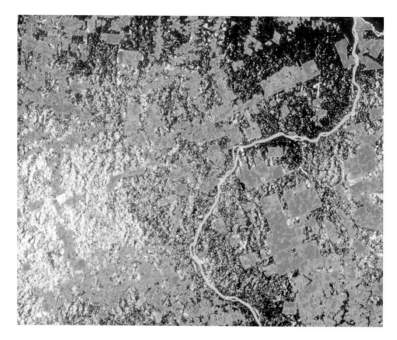

Figure 4.12 Teles Pires River area, Mato Grosso State, Brazil. The SAR sensor is very good at showing differences in land surface texture and roughness. This image shows how the difference between the rough irregular texture of tropical rainforest canopy contrasts with the relatively smoother areas of bare soil and pasture of the deforested agricultural areas. Copyright ESA 1991, 1992, original data distributed by Eurimage.

aerial photography remains an important component of remote sensing. The properties of aerial photographs were considered in Chapter 3. The applications of aerial photography in the visible and near-infrared range of the electromagnetic spectrum generally relate to situations where high-resolution information is required. The resolution may vary but is typically in the order of 0.5–1 m. This allows small individual targets to be identified. For example, individual trees can be observed to line each side of the street on the oblique aerial photograph shown in Plate 2.2.

In general, aerial photography is not useful in meteorological applications such as weather forecasting, where images of large areas are required. However, it is readily applicable to monitoring the effects of extreme weather such as flooding episodes. Such events may occur rapidly and with little warning but aerial surveying can be instigated equally rapidly. Figure 3.8 shows an aerial image obtained within the

Figure 4.13 Oil slick from a Japenese oil tanker, 11 January 1997, from the RADERSAT SAR sensor. Image courtesy of RSI, 1997.

thermal infrared which was used to pinpoint the source of forest fires in Arizona. The fine detail displayed on aerial photographs makes them particularly useful in a number of disciplines. Geologists are concerned with mapping geological structures and also discriminating rock types. Although different rock types are difficult to distinguish on black and white photographs, they are often associated with different tones, drainage networks and types and density of vegetation, all of which may be observed on large-scale photographs.

Aerial photographs viewed stereoscopically can provide information on slope variability, which is an important parameter in evaluating potential new road layouts. Landscape evaluation is easily undertaken from an aerial perspective. Figure 4.14 shows two areas separated by 10 km which are underlain by the same rock type, though an examination of the photographs illustrates that they have very different characteristics. Figure 4.14a shows this is a relatively fertile region characterised by regular field patterns, farms and roads. However, there is no discernible habitation or vegetation on Figure 4.14b. Both areas have been glaciated in the last 2 million years, but while Figure 4.14a represents an area of deposition, Figure 4.14b is a site of erosion.

One particular field of science in which aerial photography has made extensive contributions over the years has been archaeology, especially in identifying and mapping ancient sites. Such sites usually have dimensions measured in tens or hundreds of metres and the high resolution afforded by aerial surveys makes such an approach very cost-effective. Sites that appear on the surface to have topographic signatures, such as earth banks or walls, are often best observed with a low illumination angle. Many archaeological sites have a regular shape such as a rectangle or circle which also makes them show up on aerial photographs (Figure 4.15a); many which are no longer apparent upon a surface investigation may, under particular conditions, be visible from the air. The soil cover over a buried wall is thinner than at other locations, and this influences the vegetation growing over it. Conversely, an ancient ditch may become infilled with material yet not be associated with any topographic features. However, drainage and possible nutrient content in the soil-filled ditch may be different from that in the rest of the field, which influences the type, density or vigour of the vegetation that grows on it. Such crop marks are often only apparent in particular situations, such as during drought years or at the beginning of the growing season. An example of a crop mark which identified an ancient site is shown in Figure 4.15b.

The reader should consult Avery and Berlin (1992) and Lillesand and Kiefer (1994), who provide numerous examples in which aerial photographs are used for a range of applications: engineering projects, forestry,

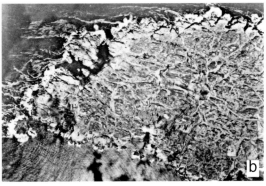

Figure 4.14 (a) Aerial photograph characterised by regular field patterns of farms and roads. (b) Nearby area showing extensive rock outcrops and no sign of habitation or cultivation. Images approximately 2 km wide. Based on the Ordnance Survey of Ireland by permission of the government (permit number 6259).

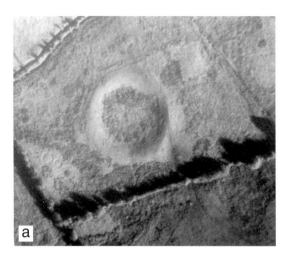

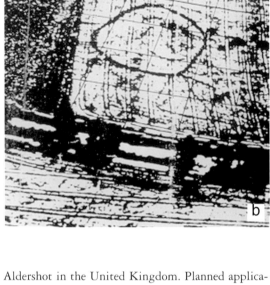

Figure 4.15 (a) Circular earthwork on the surface showing the presence of an archaeological stucture. Image courtesy of the Discovery Programme. (b) An example of a crop mark (shown as a dark ring) which identifies an ancient site. Photograph courtesy of Department of Archaeology, National University of Ireland, Cork.

soils, agriculture, land use, geology, water resources and urban planning.

Applications Using High Spatial Resolution Satellites (10 m Spatial Resolution)

The era of high spatial resolution imaging of the Earth's surface for civilian applications began in 1991 when Interbranch Association Sovinformsputnik was founded by a number of enterprises in the Russian defence industry. Its main purpose is the commercial distribution of Russian high-resolution optical satellite data acquired during previous missions and to plan new high-resolution satellite programmes under licence from the Russian Space Agency. The main sensor package comprises a topographic camera (TK-350) and the high-resolution KVR-1,000 camera flown together on a Cosmos spacecraft. Panchromatic photographic images are produced with a ground resolution of 10 m for the TK-350 camera and 2 m for the KVR-1,000 camera. Figure 4.16 shows a typical KVR-1,000 panchromatic photograph of part of Farnborough and

Aldershot in the United Kingdom. Planned applications for high spatial resolution satellites include city planning, zoning support, additional detailed information to add to existing land bases, transportation and development planning, property marketing, cadastral mapping support, vegetation community mapping and pollution event monitoring. A number of systems currently provide (or will soon provide) high spatial resolution satellite imagery (Table 4.4).

The high-resolution small satellites (Smallsats) are designed to be small and light to launch, with single-sensor payloads. The first very high-resolution small satellite Earlybird was launched by Earthwatch in December 1997 but failed soon after launch.

The choice of high spatial resolution images will be much wider than for government- and state-operated medium- and low-resolution systems, which should provide better access and availability, possibly at cheaper prices as a result of the keen competition between commercial providers. Improved local and portable data capture and processing technology which has already been proved in the research sector, combined with improved Inter-

Table 4.4 High-resolution satellites and sensors (10m or better spatial resolution) which have been recently launched or which are due for launch

Satellite system	Orbit altitude	Spatial resolution	Proposed launch date and other information
SPOT 5	832 km	10 m or 20 m XS, 2–3 m PAN	2002
IRS-1D	817 km	5.8 m PAN, 23 m or 70 m XS	Launched 1997
SPIN-2	220 km	2 m or 10 m PAN	Periodic
IKONUS	680 km	1 m PAN, 4 m XS	1999 (failed)
Orbview-3	470 km	1 m PAN, 4m XS	1999
Earlybird	472 km	3 m PAN, 15 m XS	Failed post-launch Dec. 1997
Quickbird	600 km	82 cm PAN, 4m XS	1999

Key: PAN = panchromatic; XS = multispectral.

net data transfer, will also drive the cost of data down, and increase availability to more users. The more affordable Smallsat technology is also likely to herald an increase in specialised single or limited-application missions with sensors tuned to specific user requirements rather than the general requirements of the multi-user technology-driven missions that we have been accustomed to since the 1960s. In the next few years, new portable satellite reception technology, which is currently under development, and the advent of powerful personal computer processing will make desktop reception available for high- and medium-resolution images as well as low-resolution images (Figure 4.17). At present, high- and medium-resolution data are expensive, more difficult to acquire and can only be received and processed using relatively high-technology, specialised ground stations with archiving facilities, or from relatively few data suppliers.

4.6 CHAPTER SUMMARY

Low-resolution data, even if not detailed enough, can act as a supplement for higher resolution surveys by filling in temporal and spatial gaps in data. Large area coverage lends its use to synoptic surveys of Earth's atmosphere and surface, and it is useful for environmental monitoring applications that require real-time surveying and constant situation updating, primarily:

● Measurements of atmospheric constituents, e.g. water vapour or ozone
● The acquisition of atmospheric temperature profiles
● Monitoring of air pollution (often with non-image satellite data)
● Nephanalysis and cloud movement/weather-system tracking
● Weather forecasting
● Measurement of wind speeds

Figure 4.16 A typical KVR-1,000 DD5 panchromatic photgraph of Farnborough and Aldershot in Hampshire. Copyright DERA, 1992.

- Monitoring of severe meteorological hazards
- Ocean currents and coastal sediment tracking
- Marine pollution events
- Seasonal vegetation status or landscape change over large areas
- Continental geology
- Natural disaster monitoring, e.g. floods, fires or volcanoes.

Medium spatial resolution data provide detailed smaller area coverage more suitable for local-area surveys of Earth's surface. Medium-resolution sensors

generally give more flexibility in spectral detail in non-visible parts of the spectrum than low-resolution systems, especially the infrared and microwave. Some uses of medium spatial resolution images include:

Geology

- Geological rock type discrimination and mapping
- Delineation of geological faults, fractures and other structural features

Figure 4.17 NRI/BURS experimental transportable high/medium-resolution imagery reception system. Image courtesy of Enviromental Sciences Department, Natural Resources Institute, University of Greenwich.

- Delineation of potential ore bodies and water-bearing rocks
- Production of geological maps of remote or poorly known regions and updating of existing maps
- Volcano monitoring
- Petroleum and gas exploration.

Land Use and Vegetation

- Landscape change detection studies
- Updating topographic, vegetation and land-use maps
- Discrimination of different crop and vegetation types
- Measurement of crop areas
- Soil type discrimination
- Determination of crop vigour or stress.

Water Applications

- Water-body pollution and sediment concentration
- Determination of water-body areas
- Measurements of algal blooms

- Mapping of sea-ice distribution
- Determination of sea-surface temperatures.

Radar Sensors

- Determination of sea-surface elevation
- Measurements of wave heights and directions
- Topographic mapping
- Sea-ice mapping and navigation
- Night and all-weather imaging.

The disadvantage of high- and medium-resolution data is the lack of frequent coverage of the same area of the Earth's surface. However, the new Smallsats are addressing this problem in missions dedicated to specific applications and side-looking capability sensors.

At present, it is expensive, more difficult to acquire and more complex to process medium-resolution satellites. However, this is changing with the advent of powerful personal computer processing and lightweight satellite reception dishes, making desktop reception available for high- and medium-resolution images. New application areas are envisaged for high-resolution satellites in the following fields:

- Urban planning
- Habitat mapping
- Traffic monitoring
- Urban map creation and updating
- Property
- Land ownership boundary definition
- Plant species studies
- Pollution and natural disaster monitoring.

The more affordable Smallsat technology is also likely to herald an increase in specialised single or limited-application missions with sensors tuned to specific user requirements rather than the general requirements of the multi-user technology-driven missions that we have been accustomed to since the 1960s. Aerial photography is of most benefit where high spatial resolution of small areas is required. Thus it has proved useful in archaeological and agricultural investigations and in urban and geological studies. Images tends to be obtained infrequently. However, they can be obtained at very short notice; aerial remote sensing has thus proved useful for monitoring natural and man-made disasters.

SELF-ASSESSMENT TEST

1   Which environmental science was first to make use of satellite remote sensing technology?

2   (a) What is nephanalysis and how does it help in rainfall monitoring?
    (b) How are satellite data used in UK Meteorological Office weather forecasts?

3   Explain why satellite data are so important in tropical storm tracking.

4   What does SST stand for and why is it useful?

5   What is a vegetation index and how can it be used in environmental monitoring?

6   What are the advantages of using low spatial resolution satellite data for geological mapping?

7   (a) Why do medium and high spatial resolution satellites have limited use for meteorological applications?
    (b) Name three non-meteorological applications of NOAA AVHRR data.

8   Recommend the most suitable types of image data for the following environmental applications: (a) severe storm tracking; (b) crop failure warning in semi-arid regions; (c) land-use mapping in the UK; (d) tropical rainforest clearance; (e) urban mapping and (f) ship navigation through icefields. Give reasons for your choices.

FURTHER READING

Avery, T. E. and Berlin, G. L. (1992) *Fundamentals of Remote Sensing and Airphoto Interpretation*, 5th edition, Englewood Cliffs NJ: Prentice-Hall. (Chapters 11, 13 and 14)

Campbell, J. B. (1996) *Introduction to Remote Sensing,* 2nd edition, London: Taylor and Francis. (Chapters 17 and 19)

Drury, S. A. (1993) *Image Interpretation in Geology*, London: Chapman and Hall. (Chapter 4)

European Space Agency (1995) *New Views of the Earth – scientific achievements of ERS-1*, Noordwijk, Netherlands: European Space Agency Publication SP-1176/1.

Harries, J. E. (1995) *Earthwatch: the climate from space*, Chichester: John Wiley and Sons in association with Praxis Publishing Ltd. (Chapter 5)

Legg, C. (1989) 'The potential of meteorological satellite data for monitoring flooding disasters', Farnborough: National Remote Sensing Centre, NRSC SP (89) WP31.

Lillesand, T. M. and Kiefer, R. W. (1994) *Remote Sensing and Image Interpretation*, New York: John Wiley and Sons. (Chapter 3)

Lo, C. P. (1986) *Applied Remote Sensing*, New York: Longman (Chapters 2 and 3)

Meteorological Office (1997) '*Nimrod: the leading edge of nowcasting*', Bracknell: Meteorological Office Factsheet 97/867.

APPENDIX A

ANSWERS TO SELF-ASSESSMENT TESTS

CHAPTER 1

1 Some advantages of remote sensing are:
   (a) the ability to view large parts of the globe at different scales;
   (b) the capability to monitor regions which may be very remote or where access is denied;
   (c) the ability to analyse different surfaces at wavelengths not detectable to the human visual system;
   (d) the ability to obtain imagery of an area at regular intervals over many years in order that changes in the landscape can be evaluated;
   (e) the capability to see human-induced effects on our planet.

2 Some disadvantages of remote sensing include:
   (a) a certain skill level is required to interpret the imagery;
   (b) any interpretation based solely on remotely sensed data should be treated with caution unless supported by ground verification data.

3 Scanner systems can operate within the visible range of the electromagnetic spectrum. However, scanner systems can also operate within some non-visible parts of the electromagnetic spectrum that cannot be accessed by photographic means.

4 The 1960s were of great importance to the science of remote sensing because that decade saw the launch of many remote sensing systems into space.

Before 1960 remote sensing involved almost exclusively airborne systems. Also, computing technology was beginning to improve sufficiently for the large amounts of data obtained from spaceborne platforms to be handled adequately.

5 Geostationary satellites operate at a height of approximately 36,000 km above the Earth.

6 Photographs are obtained by a camera system on a film by recording electromagnetic radiation within the ultraviolet to photographic infrared range of the spectrum. Images can be acquired digitally by scanner systems in the visible and photographic infrared (similar to photographs) but also in the thermal and microwave parts of the electromagnetic spectrum.

CHAPTER 2

1 Using the formula given in equation 2.1 and re-arranging gives:

$$\lambda = \frac{c}{f}$$

$$= \frac{3 \times 10^8}{5 \times 10^{10}}$$

$$= 0.6 \times 10^{-2} \, m$$

$$= 0.6 \, cm$$

Electromagnetic radiation with a wavelength of

0.6 cm is within the microwave part of the electromagnetic spectrum (see Figure 2.2).

2 The colour or response of a water body is determined by reflectance and scattering from within the body itself. Water absorbs virtually 100 per cent of infrared radiation; none is therefore available for reflection and scattering, resulting in a black signature. The shorter blue/green wavelengths penetrate clear water the farthest and are thus available for reflection and scattering.

3 Selective scattering is wavelength dependent (and inversely proportional to wavelength) whilst non-selective scattering affects electromagnetic radiation of all wavelengths equally. The two forms of selective scattering are Rayleigh and Mie scattering. Rayleigh scattering is inversely proportion to the fourth power of wavelength while Mie scattering is inversely proportional to wavelength.

4 A false colour image is one in which the observed colours are different from the colours of the scene as observed by the human visual system. It is produced by projecting and co-registering three images obtained in separate wavebands by colours which correspond to other wavebands. A standard false colour image is produced by projecting and co-registering an image obtained within the green range in blue light, an image obtained within the red range in green light and an image obtained within the infrared in red light.

5 Water vapour absorbs radiation with wavelengths of 1.4 µm, 2.7 µm and 6.3 µm, oxygen absorbs radiation at 6.3 µm and carbon dioxide absorbs radiation at 2.7 µm.

6 First of all calculate how far the buildings are apart from the map. The scale of the map is 1:30,000, therefore 1 mm on the map is equivalent to 30,000 mm = 30 m. As the buildings are separated by 54 mm on the map, then in reality they are separated by $30 \times 54 = 1,620$ m. This

distance is equivalent to 85 mm on the aerial photograph. Therefore:

$$85 \, mm = 1,620 \, m = 1,620,000 \, mm$$

$$1 \, mm = 1,620,000/85$$

$$= 19,058$$

Therefore the scale is 1: 19,058.

A road which is seen to be 15 mm long on the photograph is 285.87 m long (15×19.058).

7 An active remote sensing system carries its own source of electromagnetic radiation with which it illuminates the scene. It thus both produces radiation and also records the radiation that is returned from the scene. A passive remote sensing system uses the Sun as its source of illumination and only records the reflected or emitted radiation coming from the scene. The most commonly used active remote sensing systems operate in the microwave part of the electromagnetic spectrum, for example radar. This is because the signal produced by natural microwaves emanating from a scene is very weak because the intensity of the Sun's radiation in the microwave range is very low (Figure 2.3).

8 The most significant aspect of the spectral vegetation in the 0.4–1.1 µm waveband is the massive increase in reflectance that occurs at the red/infrared boundary, which is known as the red edge – see Figure 2.11.

CHAPTER 3

1 For the television masts H = 1,700 m (2,500 − 800) and substituting into equation 3.1 gives a scale of 1:15,111. Therefore 1 mm on the photograph is equivalent to 15.1 m, and 23 mm is equivalent to a ground separation of 348 m. Similarly for the buildings with a value of H of

2,300 m (2,500 − 200), the scale is equal to 1:20,444. Therefore they are separated by a distance of 470 m.

2 Both Landsat 1 and Landsat 5 were put into near polar sun-synchronous orbits. However, Landsat 1 was at an altitude of 918 km while Landsat 5 was at a height of 705 km. Landsat 5 has the shorter period and has a repeat cycle of 16 days whereas the repeat cycle for Landsat 1 was 18 days.

3 SPOT is a pushbroom system which has a better spatial resolution than Landsat. In multispectral mode the resolution is 20 m and in panchromatic mode it is 10 m. The spatial resolution for Landsat is 30 m. SPOT has a steerable mirror which allows it to obtain imagery to the side of and not just immediately below the satellite. This allows stereoscopic images to be produced for SPOT; tilting the mirror may improve the repeat cycle.

4 The completed table is shown below.

Wavelength (cm)	Depression angle (degrees)	Radar rough (cm)	Radar smooth (cm)
3	45	> 0.96	< 0.17
6.6	30	> 3.0	< 0.528
23	60	> 6.04	< 1.062

Using equations 3.8 and 3.9, a body is radar rough for a system with a wavelength of 3 cm and a depression angle of 45 degrees if:

$$H > \frac{3}{4.4 \sin 45} > \frac{3}{3.11} > 0.96 \, cm$$

and radar smooth if:

$$H < \frac{3}{25 \sin 45} < \frac{3}{17.67} < 0.17 \, cm$$

For a wavelength of 6.6 cm and a radar-rough value of 3 cm, using equation 3.8 gives:

$$3 > \frac{6.6}{4.4 \sin \gamma}$$

thus $\sin \gamma = 6.6/13.2 = 0.5$ therefore $\gamma = 30$ degrees.

Substituting this depression angle into the radar-smooth equation gives a value of 0.528 cm.

Using the radar-smooth equation with a wavelength of 23 cm gives:

$$1.062 < \frac{23}{25 \sin \gamma}$$

therefore $\sin \gamma = 23/26.55 = 0.868$, therefore $\gamma = 60$ degrees. From the radar-rough equation $H > 6.04$ cm.

5 The depression angle (γ) for a radar system is the angle measured from the horizontal to the line joining the transmitting antenna and the object imaged (see Figure 3.21). The look angle is the angle between the vertical and the line joining the antenna to the target. It is thus equal to 90 degree minus the depression angle.

6 GOES obtains measurements at 6.7 μm, which is not an atmospheric window. Water vapour absorbs at this wavelength and the purpose of the measurements is to obtain data on atmospheric water vapour concentrations.

CHAPTER 4

1 Satellite images have been used in meteorology for weather forecasting since 1960.

2(a) Nephanalysis is the analysis of cloud imagery to determine the amount of cloud, its pattern and type, which in turn gives information on the meteorological conditions and weather systems associated with certain cloud formations and characteristics. It is an important element of weather forecasting using satellite imagery. Cloud types and formations typical of rainfall can be identified and the temperature of clouds can be measured by thermal infrared imagery to determine whether the cloud is at the right temperature for its environmental location to produce rainfall. Using a series of sequential images of this type enables weather forecasters to monitor rainfall.

(b) Visible and infrared satellite imagery from both Meteosat and NOAA AVHRR satellites is automatically analysed by the UK Meteorological Office and combined with radar and rain-gauge data analyses in their Nimrod system, in order to produce more accurate and broad-scale forecasts.

3 Conventional records can often miss tropical storms as the compilation of these records relies entirely upon sightings from aeroplanes, coastal weather stations and ships, which are often not in the right location at the right time to observe every storm that occurs.

4 SST stands for Sea Surface Temperature and it is useful for determining the location of thermal pollution, warm ocean currents and likely locations for shoals of fish.

5 A vegetation index is a measure of vegetation greenness or health and has been developed as a method of enhancing healthy vegetation on satellite images, which is an important environmental parameter in arid and semi-arid regions.

6 Low spatial resolution imagery is advantageous to geological mapping as it covers large areas, thus allowing geologists to see regional and continental-scale structures. The images are also much cheaper to purchase and process than high spatial resolution data.

7(a) Medium and high spatial resolution satellites have limited use for meteorological applica-

tions as the areas covered by single images of this resolution, being around 100–200 km square, are not large enough to cover complete weather systems. It is therefore difficult to analyse these events. Imaging intervals are not frequent enough to monitor adequately the dynamic changes in weather systems on an hourly basis.

(b) NOAA AVHRR imagery can be used for vegetation health monitoring, sea-surface temperature/ocean current determination, snow mapping, straw or bush fire monitoring, broad-area geological mapping, crop monitoring and pollution detection.

8(a) Low spatial resolution meteorological satellite data in both the visible and thermal infrared, obtained every few hours for the duration of the storm event.

(b) Low spatial resolution data from NOAA AVHRR which records image data in both the visible and near infrared so that vegetation health can be determined by ratioing the visible and near infrared image bands. The low spatial resolution enables a broad-area view of the stressed vegetation resulting from drought or disease so that governmental and disaster warning agencies can issue warnings of the extent of the problem and likelihood of crop failure.

(c) Medium and high spatial resolution data are most suitable for land-use mapping in the UK as field sizes are relatively small compared to those in central Europe, Asia and North America.

(d) Tropical rainforest clearance is best detected using medium spatial resolution data (such as Landsat TM or MSS, or SPOT), if detail of the sizes and land uses of the cleared areas is required. Radar images are particularly useful if monitoring is important as radar wavelengths can penetrate cloud cover and image at night. Landsat TM thermal infrared and low spatial resolution NOAA AVHRR data are useful for detecting forest and bush fires, which are often used in forest clearance.

(e) High spatial resolution imagery of 10 m or finer resolution is most suitable for urban mapping. This type of imagery is currently available in panchromatic visible wavelengths from the SPOT satellites, IRS and aerial photography. However, in the near future higher spatial resolution (80 cm to 2 m) satellite data will be available from new small satellites, which will greatly improve the use and application of satellite images to urban land use and mapping studies.

(f) Synthetic aperture radar images are the most suitable data for ship navigation through ice-fields, as they provide day and night imaging in all weather conditions. The detection of ice texture enables the analyst to determine whether the ice covers water or land and gives an indication of its thickness and state.

APPENDIX B

ACRONYMS USED IN REMOTE SENSING

A bewildering array of acronyms exists in remote sensing and one could probably write a book on these alone. The following list indicates the main ones that have been mentioned in the text or which may be encountered in other remote sensing literature.

ADEOS	Advanced Earth Observing Satellite
AMI	Active Microwave Instrument
AMPS	Airborne Multisensor Pod System
ARTEMIS	African Real Time Environment Data and Information Service
ATSR	Along-Track Scanning Radiometer
AVHRR	Advanced Very High Resolution Radiometer
BIAS	Bristol-NOAA InterActive rainfall-monitoring Scheme
BNSC	British National Space Centre
BURS	Bradford University Remote Sensing Ltd
CCD	Charge-coupled device
CCRS	Canadian Centre for Remote Sensing
CCT	Computer Compatible Tape
CD-ROM	Compact Disk Read Only Memory
CEC	Commission of European Countries/Communities
CEO	Centre for Earth Observation
CODE	Coastal Ocean Dynamics Experiment
CZCS	Coastal Zone Colour Scanner
DERA	Defense Evaluation and Research Agency
DMSP	Defense Meteorological Satellite Program
EOSAT	Earth Observation Satellite Company
EROS	Earth Resources Observation System

ERS	European Remote Sensing Satellite
ESA	European Space Agency
EUMETSAT	European Meteorological Satellite operations centre
FCC	False Colour Composite
FRONTIERS	Forecasting Rain Optimised using New Techniques of Interactively Enhanced Radar and Satellite data (former UK Meteorological Office weather forecasting system)
GAC	Global Area Coverage
GARP	Global Atmospheric Research Programme
GATE	GARP Atlantic Tropical Experiment
GCOS	Global Climate Observing System
GIS	Geographical Information System
GMS	Geostationary Meteorological Satellite
GOES	Geostationary Operational Environmental Satellite
GOME	Global Ozone Monitoring Experiment
GPS	Global Positioning System
HCMM	Heat Capacity Mapping Mission
HIRS	High Resolution Infrared Radiation Sounder
ICSU	International Council of Scientific Unions
IFOV	Instantaneous Field Of View
IJRS	International Journal of Remote Sensing
IOC	Intergovernmental Oceanographic Commission

IR	Infrared		SIR	Shuttle Imaging Radar
IRS	Indian Remote Sensing satellite		SMS	Synchronous Meteorological Satellite
JERS	Japanese Environmental Remote Sensing satellite		SPOT	Système Pour l'Observation de la Terre
JPL	Jet Propulsion Laboratory		SSM/I	Special Sensor Microwave Imager
JSFC	Johnston Space Flight Centre		SSM/T-1	Special Sensor Microwave Temperature Sounder
LAC	Local Area Coverage			
LFC	Large Format Camera		SSM/T-2	Special Sensor Microwave Water Vapour Sounder
MEIS	Multispectral Electro-optical Imaging Scanner			
MIR	Mid Infrared		SST	Sea Surface Temperature
MOS	Marine Observation Satellite		TDRS	Tracking and Data Relay Satellite
MSS	Multispectral Scanner			
MSSL	Mullard Space Science Laboratory of University College London		TIMS	Thermal Infrared Multispectral Scanner
MTF	Modulation Transfer Function		TIR	Thermal Infrared
MTPE	Mission To Planet Earth		TIROS	Television Infrared Operational Satellite
NASA	National Aeronautics and Space Administration			
NASDA	National Space Development Agency		TM	Thematic Mapper
			TOMS	Total Ozone Mapping System
NCDC	National Climate Data Centre		TOVS	TIROS Operational Vertical Sounder
NDVI	Normalised Difference Vegetation Index			
			TRMM	Tropical Rainfall Mapping Mission
NOAA	National Oceanic and Atmospheric Administration		UARS	Upper Atmosphere Research Satellite
NRI	Natural Resources Institute of the University of Greenwich, UK		UNEP	United Nations Environmental Project, based in Nairobi, Kenya
NRSC	National Remote Sensing Centre		UNESCO	United Nations Educational, Scientific and Cultural Organisation
NSIDC	National Snow and Ice Data Centre			
OCTS	Ocean Colour and Temperature Scanner		UN FAO	United Nations Food and Agricultural Organisation
OLS	Operational Linescan System		UV	Ultraviolet
PDUS	Primary Data User Station		VI	Vegetation Index
RADAR	RADio And Ranging		VIS	VISible waveband/s of satellite imagery
RAR	Real Aperture Radar			
RBV	Return Beam Vidicon		VISSR	Visible Infrared Spin-Scan Radiometer
RGB	Red Green Blue colour composite image			
			WCRP	World Climate Research Programme
SAR	Synthetic Aperture Radar		WMO	World Meteorological Organisation
SeaWiFS	Sea-viewing Wide Field-of-view Sensor		WV	Water Vapour channel from GMS
			WWW	World Wide Web

APPENDIX C

GLOSSARY OF REMOTE SENSING TERMS

This glossary provides a brief explanation of the main terms used in remote sensing.

Absorption band: The wavelengths of electromagnetic radiation that are absorbed by a substance. Some components of the atmosphere have a number of absorption bands. Water vapour in the atmosphere, for example, absorbs radiation with a wavelength of 6 μm.

Across-track scanning system: See transverse scanning system.

Active microwave instrument: Radar systems carried aboard the European Radar Satellites comprising a SAR and wind scatterometer.

Active remote sensing system: A remote sensing system that provides its own source of electromagnetic radiation to illuminate the target. Radar is the commonest active system employed in remote sensing. See also passive remote sensing.

Additive primary colours: The colours red, green and blue which, when added in different combinations, can produce all the colours.

Advanced Very High Resolution Radiometer: See AVHRR.

Aerial photograph: A representation of a scene recorded on photographic film from the air. Acquisition platform is usually an aircraft. See also oblique aerial photograph and vertical aerial photograph.

Aggregation: The merging of two or more separate classes into a single class. This may be used when there is a substantial overlap between classes.

Albedo: The albedo of a surface is the ratio of the radiation reflected from it to the total electromagnetic radiation incident on it.

Algorithm: The term applied to a series of instructions given to a computer.

ALMAZ: A former Russian radar satellite, mission life 1991–3.

Along-Track Scanning Radiometer: A scanning system carried aboard the European Radar Satellites. On ERS-2, it comprises a seven-channel radiometer and a microwave sounder.

Along-track scanning system: See pushbroom system.

Altimetry: Technique for the measurement of the height of a satellite above the surface. The length of time a pulse of electromagnetic radiation takes on a round trip from the satellite to the ground and back is measured and from this the height can be accurately determined. When employed over oceans, the sea height can be measured.

Amplitude: Characteristic of an electromagnetic wave. It is the distance from peak to trough.

Atmospheric window: Atmospheric windows represent the wavelengths of electromagnetic radiation that are not absorbed by components of the atmosphere. The waveband 0.4–0.7 μm (visible) is an atmospheric window. remote sensing systems are generally designed to operate within atmospheric windows.

AVHRR: Advanced Very High Resolution Radiometer. Multispectral scanner system aboard the TIROS satellites.

AVIRIS: Acronym for Airborne Visible Infrared Imaging Spectrometer. This is a remote sensing system which typically obtains data in many bands (> 200) with very high spectral resoltion.

Azimuth direction: A term used in radar to describe the direction in which the airborne sensor is moving. See also range direction.

Azimuth resolution: A term used in radar to describe the resolution in the direction in which the sensor platform is moving. See also range resolution and synthetic aperture radar.

Backscatter: A term used in radar to describe the energy that is scattered back to the airborne or spaceborne platform.

Band: A wavelength range measured by a remote sensing system. Often used to designate how many 'layers' of data are recorded by a system, e.g. Landsat TM is a seven-band system. Note that the term 'channel' is occasionally used instead of band.

Band interleaved by line: A term that describes the format in which the digital data on a computer-compatible tape are arranged. In band interleaved by line format (BIL) for an n-band system, the first n lines of data hold the digital numbers for the first line for all n bands.

Band interleaved by pixel: A term that describes the format in which the digital data for a scene are arranged. In this format, for an n-band system, the first n numbers are the DNs for the first pixel on the first line, the second set of n numbers are the n digital numbers for pixel 2 line 1 and so on.

Band ratioing: An enhancement procedure in which a new image is created by dividing the DN in one band by the DN in another band for every pixel in the dataset. Band ratioing has particular applications in mineral exploration and in vegetation surveys. Also see vegetation index.

Binary number: Numerical system based on the number 2.

Bit: Abbreviation for binary digit; can be represented by zero or one. The term is analogous to quantum and represents the smallest discrete package of information. Landsat TM, for example, transmits its data at 85 megabits per second.

Blackbody: A theoretical body which absorbs the entire radiation incident on it and which obeys the Stefan–Boltzmann Law.

Brightness value: A term synonymous with digital number.

Byte: A collection of eight bits (256 levels).

C: Letter assigned to the speed of light (and all electromagnetic radiation) which is equal to 3×10^8 m/s in a vacuum.

Camera: A device for recording the reflectance characteristics of a scene on a sensitised medium which generally operates within the visible or photographic infrared sections of the electromagnetic spectrum.

CCD: Charge-coupled device which is used as a detector on pushbroom systems.

CCT: Computer-compatible tape on which satellite data are stored.

CD-ROM: Compact Disk, Read Only Memory. An optical storage device which is commonly used to hold image data because such devices can store up to 600 Mb of data.

Central Processing Unit (CPU): Part of the computer system that processes the data according to the operator's instructions.

Classification: The procedure by which an n-band dataset is converted into a thematic image with a finite number of classes. See also unsupervised classification and supervised classification.

Coastal Zone Colour Scanner: A scanning system primarily designed to measure chlorophyll variations in the oceans which was carried onboard Nimbus 7.

Complementary colour: The complementary colour of a primary colour is produced by mixing the two remaining primary colours. Thus yellow is the complement of blue because it is produced by green and red light.

Cone: Detectors in the human eye that are sensitive to colour. See also rod.

Contrast stretch: The procedure by which the input DN range for an image, which usually spans only a small part of the available DN, is increased to encompass a larger part of the available DN. This mathematical procedure increases the contrast of the image.

Co-registration: The alignment of different bands of data or data obtained from different systems so that the row and column co-ordinates for all pixels for all bands are identical, where each pixel represents a discrete ground area.

Corner reflector: A term used in radar to describe intersecting surfaces which return the incident energy back towards the sensor and thus produce a bright signature.

Correlation: In remote sensing, it is a measure of the degree to which the DN for one band can be predicted if the DN of another band is known.

Covariance: The joint variation of two variables about their common mean. The covariance is an important statistic used in some enhancement procedures such as principal components analysis.

Cross-polarised: A term applicable to a radar system which transmits in the vertical (or horizontal) direction and measures the returned signal in the horizontal (or vertical) plane. See also parallel-polarised.

CZCS: See Coastal Zone Colour Scanner.

DARTCOM: A UK company that has developed and markets a PC-based low spatial resolution satellite data reception system.

DD5: Russian high-resolution space photography system.

Depression angle: A term used in radar to describe the angle between a horizontal plane and the direct line from transmitter to the object that is being imaged. It decreases from near to far range.

Dielectric constant: A term used in radar that is a measure of the electrical properties of a substance.

Digital elevation model: A regular grid array of numbers in which the numbers represent elevation.

Digital image: An image obtained by a scanner system in which the measured parameter (such as reflectance) is represented by an array of pixels, each of which is associated with a digital number.

Digital image processing: The manipulation of digital data by computer programs in order to improve the appearance of an image.

Digital number: See DN.

Display monitor: Component of a digital image processing system on which the image is shown.

DN: Digital number. The number assigned to a pixel which is related to the parameter being measured by a remote sensing system. For example, a low DN may represent a low reflectance in a particular waveband and a high DN a high reflectance. DN values must be integers and common ranges used are 0–63, 0–128 and 0–255. Note: for an n-band dataset each pixel is associated with n digital numbers.

DN histogram: See frequency histogram.

Doppler shift: The apparent change in frequency as a wave moves past an observer. SAR systems employ this effect when recording data in order to produce a high azimuth resolution.

Dwell time: For a pushbroom system the length of time for which a ground resolution cell is viewed by a detector. In a transverse scanning system, it is the time taken for the ground resolution cell to be swept by a scanning mirror.

Edge: An abrupt change in tone observed on an image.

Eigenvalue: A term used in principal components analysis that is a measure of the amount of variance in each principal component.

Eigenvector: A parameter that gives the orientation of a principal component axis.

Electromagnetic radiation: Energy transmitted at the speed of light by oscillating electric waves.

Electromagnetic spectrum: The range of wavelengths or frequencies of electromagnetic radiation.

Emissivity (ε): The ratio of the energy actually emitted by unit area of a surface in unit time at a given temperature to the energy emitted by unit area of a blackbody in unit time at the same temperature.

Emittance: See exitance.

EOSAT: US satellite data supplier and operating company.

ERDAS: US company that develops and sells image processing software.

ERMAPPER: Image processing software package

developed by an Australian company, Earth Resources Mapping.

ESRIN: European satellite data collection and archiving centre.

EURIMAGE: European satellite data distribution centre/agency.

Exitance: Radiant flux of energy emitted by a body. Usually stated per unit area. Can be determined from the Stefan–Boltzmann Law.

False colour composite: An image whose colours do not accord with those that would be seen with the human eye. False colour composites usually include at least one input band that is invisible to our eyes. Projecting the green range in blue, the red range in green and the infrared range in red forms a standard false colour composite.

False colour image: See false colour composite.

Far range: A term used in radar to describe that part of the image which is farthest from the transmitting antenna. See also near range and depression angle.

Feature–space plot: A graph showing the distribution of digital numbers in a dataset in which the axes are formed by different spectral bands.

Fiducial: Tick marks on the edges or corners of aerial photographs that can be used to determine the principal point of the photograph.

Filter: A mathematical procedure that alters the digital numbers in an image. See also high-pass and low-pass filter.

Foreshortening: A term employed in radar to describe the compression effect observed on slopes being illuminated by a radar beam. Also see layover.

Frequency: A property of an electromagnetic wave that is a measure of the number of wave crests passing a point in unit time. It is inversely proportional to wavelength.

Frequency histogram: A bar chart representation of the DN variation in an image. It is produced by plotting DN values (0–255) on the horizontal axis against frequency of occurrence (vertical axis). Each band will have a different frequency histogram.

Geographical Information System: A system incorporating a collection of spatially correlated datasets and a collection of computer programs with which to analyse the data.

Geometric rectification: The process by which a distorted image is corrected, usually in such a way that north is at the top of the image and the scale is constant throughout the image.

Geostationary orbit: An orbit with a period of one solar day. The satellite thus always appears to remain stationary in the sky and is at a height of 36,000 km above the Earth.

Global Ozone Monitoring Experiment: See GOME.

GOME: Global Ozone Monitoring Experiment. ERS-2 carries a spectrometer which takes measurements between 0.24 μm and 0.79 μm which allow the concentrations of atmospheric constituents such as ozone, nitrogen oxide and nitrogen dioxide to be determined.

Ground-collected verification data: A term employed in remote sensing to describe the validation of a signature obtained by a remote sensing system by an inspection of the imaged site. (Some authors refer to this as 'ground truth'.)

Ground control point (GCP): A point that can be located on both an image and a map which is used in the process of rectification. Such points are used to produce the mapping algorithms which change a distorted image into a geometrically correct one.

High-pass filter: A filter that accentuates edges and sharpens up an image. See also low-pass filter.

Histogram equalisation stretch: A non-linear contrast stretch which stretches the data depending on the number of DNs in each DN bin.

HRV: Acronym for High Resolution Visible that refers to the two sensors carried by SPOT.

Hybrid ratio image: A colour image in which, for example, two input images may be ratio images while the third input is an image obtained directly by the satellite. Thus, if the ratio of TM 1 to TM 2 is projected in red, the ratio of TM 3 to TM 4 is projected in green and TM 5 is projected in blue, the result is a hybrid ratio image. Such images may be useful as they re-introduce topographic expression into the display.

IFOV: See Instantaneous Field Of View.

IHS: An alternative means of visualising a colour image which does not use the conventional red, green, blue axes. The image is considered in terms of its intensity, hue and saturation. It is possible to convert a RGB image into an IHS image.

Illumination angle: The angle that the radiation from the Sun makes with the Earth as measured from the horizontal. This varies throughout the year, being less in the winter compared with the summer for the northern hemisphere. There is a lower annual variation near the equator.

Illumination direction: The direction from which the illumination comes when an area is being imaged. Satellite systems such as Landsat obtain their data in the morning when the Sun's illumination is from the southeast in the northern hemisphere. The illumination direction for active systems depends on the transmitter's orientation.

Image: A general term used to describe the representation of a scene obtained either photographically or digitally.

Image enhancement: The process by which data displayed on an image can be made more obvious to the human visual system.

Incidence angle: The angle that electromagnetic radiation makes with the surface measured from the vertical. This term is often applied to radar systems.

Instantaneous field of view: The solid angle through which a detector on board a remote sensing system obtains data.

Intensity Hue Saturation: see IHS.

Interferometry: The process by which a three-dimensional surface can be created by using phase differences in the returned signals for active remote sensing systems.

Kinetic temperature: Temperature of a body measured by a thermometer. See also radiant temperature.

KVR-1000: Russian space panchromatic camera with 2 m spatial resolution.

Lambertian reflector: A surface that is rough compared with the wavelength of the radiation incident on it. The incident energy is consequently scattered in all directions.

Landsat: A series of Earth-observing satellites, the first of which was launched by the United States in 1972.

Layover: A term used in radar to describe the effect of the tops of tall features being closer to the radar system than the base of the features. Also see foreshortening.

LIDAR: Acronym for Light Intensity Detection and Ranging. LIDAR remote sensing systems are active and employ lasers to map topographic variations accurately or to identify particular substances.

Lineament: A linear, topographic, tonal or textural feature observed on an image.

Line banding: A defect observed on some images in which a regular banding effect is observed where lines of data are consistently darker or lighter than the rest of the scene. It appears on multi-detector systems in which a detector's response no longer matches the other detectors on board the platform. See also sixth-line banding.

Line dropout: The loss of a line of data. This defect can be caused by a loss of signal or an error in the recording. DN values are zero for such a line. An image with a number of line dropouts may be visually improved by suitable image processing.

Logarithmic stretch: A non-linear stretch that preferentially stretches the darker parts of a scene (lower digital numbers).

Look angle: A term used in radar that is equal to 90 degrees minus the depression angle.

Low-pass filter: A filter that accentuates the low-frequency component of an image, generally 'blurring' the image.

Map projection: Process by which the elliptical Earth is represented on a two-dimensional surface.

Maximum likelihood classification: Classification procedure based on the assumption that training area datasets have a normal distribution, which involves the construction of probability contours.

Mean: A statistical measure of the average of a collection of numbers. It may be obtained by adding the numerical values and dividing by the number of measurements.

Median: The numerical value that divides a DN distribution into two equal halves.

Meteosat: Geostationary satellite that obtains data for Africa and Europe.

Microwave: Part of the electromagnetic spectrum ranging from 0.1 cm to 1 m.

Minimum distance classification: Classification procedure by which an unknown pixel is assigned to the class to which it is nearest, i.e. shortest distance to the mean of the class.

MK-4: Russian high spatial resolution space camera.

Mode: A term used in statistics that refers to the numerical value which occurs most often in a range of numbers.

Mosaic: A number of aerial photographs or satellite images joined together in order to display a larger area.

MSS (Multispectral scanner): A system that obtains data in a number of wavebands simultaneously. The MSS system may be carried by an airborne or spaceborne platform. The Landsat multispectral system operates at four wavebands.

Multispectral imagery: Imagery of an area that has been recorded in a number of wavebands. Landsat MSS, for example, records four bands.

Multispectral scanner: See MSS.

Multitemporal imagery: Imagery of the same area that has been obtained on different dates. See also multispectral imagery.

Near range: A term used in radar to describe that part of the image which is nearest to the transmitting antenna. See also far range and depression angle.

NOAA: Acronym for National Oceanic and Atmospheric Administration. This organisation maintains a number of meteorological satellites.

Non-linear stretch: A contrast stretch that stretches DN ranges by different amounts. Also see his-togram equalisation, logarithmic and power-law stretch.

Normalised Difference Vegetation Index (NDVI): A ratio image used in vegetation studies produced by subtracting the DNs for the red band from the infrared band and dividing the result by the addition of the red and the infrareds, i.e. $(IR - R)/(IR + R)$. See also vegetation index.

Oblique aerial photograph: An aerial photograph obtained by a tilted camera system.

Parallelepiped classification: Classification procedure in which an n-dimensional box encompasses the training area and unknown pixels that fall within the box are assigned to the training-area class.

Parallel-polarised: A term applicable to a radar system which measures the returned signal in the same plane as it transmits the microwave radiation. See also cross-polarised.

Passive remote sensing: Remote sensing which uses the radiation from the Sun as the source of illumination. See also active remote sensing.

Photographic infrared: The part of the electromagnetic spectrum from 0.7 μm to 0.9 μm.

Pixel: Picture element; the smallest definable unit on a digital image.

Power-law stretch: A non-linear stretch that preferentially stretches the brighter parts of a scene (higher digital numbers).

Principal components analysis: Mathematical transform that produces new images based on the variance of the multispectral dataset.

Principal point: Centre point of an aerial photograph.

Pseudocolour image: A single-band black and white image in which each digital number or range of digital numbers is assigned a colour.

Pushbroom system: Imaging system which employs an array of many small detectors, each of which measures the reflectance for an individual ground sampling cell. SPOT uses a pushbroom imaging system.

Radar: Acronym for RAdio Detection And Ranging. It is the principal form of active remote sensing and operates within the microwave part of the electromagnetic spectrum.

Radar range: See near and far range.

Radar rough: A term used in radar to describe a surface which backscatters a substantial amount of the incident energy towards the antenna. A radar-rough surface will appear bright on an image. See also radar smooth.

Radar smooth: A term used in radar to describe a surface which backscatters very little of the incident energy towards the antenna. A radar-smooth surface will appear dark on an image.

Radiance: The power emitted by a body per unit area per unit steradian.

Radiant temperature: Temperature of a body measured by a radiometer. For a real body (as opposed to a theoretical blackbody) the radiant temperature is always less than the kinetic temperature.

Radiometric resolution: The number of grey levels measured by a digital system. An 8-bit system measures 256 grey levels; the available digital numbers thus range from 0 to 255.

RAR: A Real Aperture Radar is a system which employs a long antenna in order to improve the azimuth resolution. See also SAR.

Ratio image: An image produced by dividing the DNs in one band by the corresponding DNs in another band. It is often used in vegetation studies.

Rayleigh scattering: See scattering.

Real Aperture Radar: See RAR.

Reflected infrared: The part of the electromagnetic spectrum ranging from 0.7 µm to 3.0 µm.

Remote sensing: The acquisition and recording of information about an object without being in direct contact with that object.

Resolution: See radiometric, spatial, spectral and temporal resolution.

Rod: Type of detector used by the human visual system which detects variations in brightness. See also cone.

Roughness: See radar roughness.

SAR: Acronym for Synthetic Aperture Radar. A SAR system employs the principles of the Doppler shift to improve resolution in the azimuth direction. See also RAR.

Satellite: In remote sensing, an unmanned spacecraft which orbits the Earth obtaining data.

Scale: The scale of an image relates the distance on the ground between two objects and the distance between the objects as measured on an image. It is generally expressed as a ratio or as a fraction. Thus a scale of 1: 10,000 means that a distance of 1 mm on an image is equivalent to 10 m.

Scattering: The random propagation of electromagnetic radiation as a result of interaction with various components of the atmosphere. Selective scattering is wavelength dependent and, depending on the relative size of the particles compared with the wavelength of the radiation, either Rayleigh or Mie scattering may occur. Rayleigh scattering is inversely proportional to the fourth power of wavelength and Mie scattering is inversely proportional to wavelength. Non-selective scattering is not dependent upon wavelength.

Scatterplot: Two (or three)-dimensional plot showing the DN distribution between bands.

SEASAT: Radar satellite launched in 1978 that was primarily designed to obtain data on the oceans.

SeaWiFS: Sea-viewing Wide Field-of-view Sensor. Satellite-based remote sensing system designed to obtain data about phytoplankton variations in the oceans.

Shuttle Imaging Radar: Radar experiments performed on board the Space Shuttle. To date - SIR-A/B/C have been carried out.

Signal-to-noise ratio: The ratio of the signal from the ground which carries information to the noise due to aberrations in the electronics or defects in the scanning system which degrade the signal to some extent.

Signature: The various elements that, combined, describe how a particular feature appears on an image. For example, a rock type may be characterised by its colour and texture.

SIR: See Shuttle Imaging Radar.

Sixth-line banding: A defect observed on some Landsat MSS images in which every sixth line is darker or lighter than the surrounding lines. It is caused by the degradation of a detector but may be rectified by suitable processing.

Slant-range: A term used in radar, the direct distance from a radar source to the target.

Space Shuttle: Reusable United States space vehicle which has been used to launch satellites and on which remote sensing experiments have been performed.

Spatial resolution: A measure of the amount of detail that can be observed on an image.

Spectral resolution: A measure of the number and width of bands obtained by a remote sensing system.

Specular reflector: A surface that is smooth compared to the wavelength of the radiation that is hitting it, so that the incident energy is reflected at the same angle as that at which it strikes the surface. Also see Lambertian reflector.

SPOT: Series of French polar-orbiting satellites. To date, four have been launched.

Standard deviation: A statistical measure of the spread or dispersion of data about the mean. Many programmes in digital image processing such as principal components analysis or classification calculate and use this parameter.

Standard false colour composite: See false colour composite.

Stereoscope: An optical device which allows the simultaneous examination of a pair of overlapping aerial photographs to produce a three-dimensional effect.

Sun-synchronous: A term applied to a satellite's orbit such that the plane of the satellite's orbit is always at the same angle to the Sun.

Supervised classification: A classification process in which an image is separated into a number of information classes based on the statistical characteristics of training areas outlined by the operator. See also unsupervised classification.

Surface scattering: A term used in radar to indicate that the main scattering occurs at the surface. See also volume scattering.

Synergistic display: A display in which images from different systems are co-registered and displayed simultaneously. For example, SPOT data may be combined with radar data. Such images may allow different surfaces to be discriminated more easily.

Synthetic Aperture Radar: See SAR.

Temporal resolution: A measure of how often an area is imaged by a particular remote sensing system. Landsat 5, for example, has a repeat cycle of 16 days.

Texture: A qualitative description of the rate of change of tone on an image.

Thematic map: A map produced by classifying an image in which a colour represents a specific theme such as land cover.

Thematic Mapper: This is a scanning system that has been carried by Landsats 4 and 5. It obtains data in seven wavebands with a 30 m resolution in bands 1–5 and 7 and 120 m resolution in band 6 (thermal).

Thermal capacity: A measure of the ability of a substance to retain heat.

Thermal conductivity: A measure of the ability of a substance to transfer heat.

Thermal inertia: A measure of how a body responds to changes in temperature.

TM: See Thematic Mapper.

Training area: An area on an image which the operator, using prior knowledge, knows belongs to one specific class which will be used in a supervised classification procedure. Thus, for example, an operator may delineate a known forest as a training area and use the generated signature for this area to search for other regions on the image which have the same DN characteristics as the training area and which may subsequently be assigned to the forest surface class.

Transverse scanner: A scanner system which employs an oscillating mirror which sweeps across the terrain in parallel strips which are at right angles to the platform movement.

Ultraviolet radiation: Part of the electromagnetic spectrum ranging from $0.03\,\mu m$ to $0.4\,\mu m$.

Unsupervised classification: An automatic classification process in which the image is divided into a number of spectral classes based on DN distribution. The operator has minimal input. See also supervised classification.

US CROPCAST: Commercial product derived from meteorological satellite data giving updates on crop status/health throughout the growing season.

Variance: A statistical parameter used to measure the spread or the dispersion about the mean. The square root of variance is the standard deviation.

Vegetation index: A ratio image used in vegetation studies produced by dividing the data in the infrared band by the DNs in the red, i.e. (IR/R). See also Normalised Difference Vegetation Index.

Vertical aerial photograph: An aerial photograph taken with the camera pointing vertically down. Such photographs are obtained for stereoscopic viewing.

Visible light: That part of the electromagnetic spectrum that can be detected by the human visual system. It extends from 0.4 μm to 0.7 μm.

Volume scattering: A term applied to radar in which the radar signals are returned after interacting with the interior of a target such as a forest canopy. See also surface scattering.

Waveband: A term used to describe a range of the electromagnetic spectrum. Thus MSS 7 obtains data in the 0.7–1.1 μm waveband.

Wavelength: Characteristic of an electromagnetic wave, being the distance from peak to peak.

Whiskbroom scanner: See transverse scanner.

X-rays: Electromagnetic radiation with a very short wavelength (approximately 10^{-10} m). It is not used in remote sensing of the Earth.

APPENDIX D

REMOTE SENSING SOURCES OF INFORMATION

The general term 'sources of information' is used in the title of this appendix because it encompasses a range of themes. One might, for example, simply require information on the satellites that are currently in operation and the characteristics of their sensors. Alternatively, one may not require such information but need to know where specific datasets can be obtained or what software is available to process the data. Many remote sensing resources exist and it would not be possible to list them all. However, it is possible to categorise them broadly into government agencies, educational facilities and private companies. It should be realised that these categories are not mutually exclusive: university-based campus companies are today quite common. The nature of the 'service' provided also varies greatly. Some centres may act as repositories for digital data which may be purchased, whereas others may be contracted to undertake specific projects or even provide digital image processing facilities which may be hired for a short duration. The addresses of companies and institutions that contributed imagery or data for volumes I and II of *Introductory Remote Sensing* are listed in Appendix E.

Many third-level institutions and universities have a remote sensing capability, even if it is quite elementary, and are usually willing to provide information on remote sensing to interested parties. Generally (though not exclusively!) remote sensing is not a separate subject but is housed in the Department of Geography, Department of Geology or the Department/School of Environmental Studies.

Various texts have been listed at the end of each chapter which readers should also consult if they wish to obtain a greater insight into remote sensing. A number of remote sensing atlases have been produced, an examination of which is useful for building up experience of the signatures of various features. Examples include:

Images of the World (1984) published by Collins and Longman.
Earthwatch: a survey of the world from space (1981) by C. Sheefield and published by Sidgewick and Jackson.
Man on Earth: the marks of man, a survey from space (1983) by C. Sheefield and published by Sidgewick and Jackson.
NASA: views of Earth (1985) by R. Kerrod and published by Admiral Books.
Britain from Space (1985) by R. K. Bullard and R. W. Dixon-Gough. Published by Taylor and Francis.
SEASAT Views North America, the Caribbean and Western Europe with Imaging Radar (1980) by J. P. Ford, R. G. Blom, M. L. Bryan, M. I. Daily, T. H. Dixon, C. Elachi and E. C. Xenos. JPL Publication 80-67.
Space Shuttle Columbia Views the World with Imaging Radar: the SIR-A experiment (1983) by J. P. Ford, J. B. Cimino and C. Elachi. JPL Publication 82-95.
Shuttle Imaging Radar Views the Earth from Challenger: the SIR-B experiment (1986) by J. P. Ford, J. B. Cimino, B. Holt and M. R. Ruzek. JPL Publication 86-10.
Thematic Mapping from Satellite Imagery (1988) by J. Denegre. Published by Elsevier Applied Science Publishers Ltd.
The Home Planet (1988) edited by K. W. Kelley and published by Queen Anne Press.

Although each chapter in this book details further readings, a difficulty with providing a 'further reading' section in a subject like remote sensing is that, because it is evolving quite rapidly, some of the references may soon be out of date. In order to be aware of the most current up-to-date research in aspects of remote sensing, it is necessary to consult the specialised academic journals which publish the results of such research. Notable journals include:

International Journal of Remote Sensing
Remote Sensing of Environment
Canadian Journal of Remote Sensing
Institute of Electrical and Electronic Engineers: Transactions on Geoscience and Remote Sensing
Photogrammetric Engineering and Remote Sensing
Sistima Terra.

In addition, there are magazines which can also provide up-to-date information such as *SPOT* magazine by SPOT IMAGE, or *Earth Observation Quarterly,* which is published by ESA. As the reader is aware, remote sensing is employed in various fields; consequently, papers which involve the use of remote sensing appear in other journals. Thus, for example, the *Geological Journal* and the *Journal of the Geological Society, London,* have at different times published papers where remote sensing has been employed in a geological context. Meteorological journals such as *Weather, Atmospheric Environment* and the *International Journal of Climatology* regularly publish papers in which the interpretation of remote sensing data is an important component. Most geographical journals include occasional papers with a remote sensing theme and journals such as *Transactions of the Institute of British Geographers, Annals of the Association of American Geographers* and *Progress in Physical Geography* should also be consulted.

Regular conferences on remote sensing take place, with an increasing trend towards specialisation. Thus one conference may concentrate on a specific theme such as applications of remote sensing for forestry whereas another may concentrate on the applications of a specific sensor, such as the ERS radar system. Information about these conferences may generally be obtained from third-level institutions or from societies such as the Remote Sensing Society, c/o Department of Geography, University of Nottingham, Nottingham, UK, or the American Society for Photogrammetry and Remote Sensing, 5410 Grosvenor Lane, Suite 210, Bethesda, MD 20814-2160.

Companies which supply digital image processing software that can be used to analyse remote sensing data include:

EASI/PACE: IS Ltd, Atlas House, Atlas Business Centre, Simonsway, Manchester M22 5HF, UK, or PCI Enterprises, Richmond Hill Enterprises, Ontario L4B 1MS, Canada.

ENVI: Floating Point Systems UK Ltd, Ash Court, 23 Rose St, Wokingham, Berkshire RG40 1XS, UK.

Dimple: Cherwell Scientific Publishing Ltd, The Magdalen Centre, Oxford Science Park, Oxford OX 4GA, UK.

ERDAS Imagine: 2801 Bufford Highway, Suite 300, Atlanta, GA 30329-2137, USA or Telford House, Fulbourn, Cambridge CB1 5HB, UK.

ERMAPPER: Earth Resource Mapping, 4370 La Jolla Village Drive, Suite 900, San Diego, CA 92122-1253, USA, or Blenheim House, Crabtree Office Village, Eversley Way, Egham, Surrey TW20 8RY, UK.

DRAGON: Goldin–Rudahl Systems Inc., 6 University Drive, Amherst, MA 01002, USA or IS Ltd, Atlas House, Atlas Business Centre, Simonsway, Manchester M22 5HF, UK.

Idrisi: Idrisi Resources Centre, Manchester Metropolitan University, Department of Environmental and Geographical Sciences, John Dalton Building, Chester Street, Manchester M1 5GD, UK, or The Idrisi Project, Clark University, Laboratories for Cartographic Technology and Geographic Analysis, 950 Main Street, Worcester, MA 01610, USA.

IGIS: LaserScan Ltd, Cambridge Science Park, Milton Road, Cambridge CB4 4FY, UK.

RSVGA: Eidetic Digital Imaging, Brentwood Bay, BC V0S 1A0, Canada.

V-image: VYSOR Integration Inc., Gatineau, Quebec J8T 5W5, Canada.

TeraVue: Geo-Services International (UK) Ltd, Unit 5, Des Roches Square, Witan Way, Witney, Oxfordshire OX8 6BE, UK.

TNTMips: Nigel Press Associates Ltd, Edenbridge, Kent TN8 6HS, UK.

One of the most extensive remote sensing resources is the World Wide Web (WWW), which consists of a networked system of information provided from a range of sources. A user in Paris can log on to data held by an organisation in New York. Search engines on the WWW allow the input of keywords in order to locate the relevant sites. However, a disciplined approach is needed to obtain specific information. The term 'remote sensing' will elicit over 300,000 references. Because of the extensive linkages between different sites on the WWW, there is no unique pathway to a specific piece of information. The following sites indicate some of the information that can be obtained from the WWW. The sites listed here (shown within square brackets) are correct at the time of writing (1998) though some may have changed address or ceased functioning since then.

The core of remote sensing is imagery and one may examine thousands of images from a number of sites. As regards radar imagery, the NASA/JPL Imaging Radar Site

[http://southport.jpl.nasa.gov]

provides superb examples of data obtained during the SIR-C mission in 1994. Currently, the imagery is divided into a number of categories (e.g. archaeology, geology, glaciers and oceanography) and each image is accompanied by a description. A facility on this site allows one to download either the image or the data. Typically an image is 300 Kilobytes in size and in a jpeg or gif format, though the full datasets are substantially larger (15–60 Megabytes). However, the raw data can then be digitally processed to enhance different features. The range of features displayed on these datasets, and the amount of data available, make it possible to produce a large number of images designed to display an extensive range of environmental and cultural features which can form the basis of a teaching programme. NASA provides an image gallery site

[http://www.nas.gov/gallery/photo/index.html]

which links to a range of other NASA sites such as the Johnston Space Flight Center, which is a repository for thousands of images obtained on manned space missions. Many of these images relate to the earliest United States missions (Mercury, Gemini and Apollo) though imagery obtained from the Skylab missions in the 1970s and Space Shuttle flights in the 1980s and 1990s are also included. Most of the images are photographs taken by the astronauts. Ocean colour data from the CZCS and SeaWiFS can be obtained at

[http: //daac.gsfc.nasa.gov/].

The Canadian Centre for Remote Sensing allows the user to observe different parts of Canada with different types of remote sensing sensors. MEIS, SPOT, TM and radar imagery is included. The site location is

[http://www.ccrs.nrcan.gc.ca/ccrs/xhomepage.html].

It is possible to view a selection of NOAA images, mostly of storms and hurricanes, at a National Climate Data Centre site:

[http://www.ncdc.noaa.gov/ol/satellite/olimages. html].

APPENDIX E

CONTRIBUTORS TO *INTRODUCTORY REMOTE SENSING*

Texts such as *Introductory Remote Sensing: Principles and Concepts* and *Introductory Remote Sensing: Digital Image Processing and Applications* rely heavily on the goodwill of many individuals and organisations to provide relevant remote sensing imagery and data and the permission to use them. We would like to thank the organisations listed below who made such contributions. It is invidious to pick out individuals who provided a great degree of assistance but especial thanks go to Martin Critchley of ERA-Maptec, Hervé Lemeunier of SPOT IMAGE and Annie Richardson of JPL who provided most of the datasets included on the CD for the companion volume.

American Society for Photogrammetry and Remote Sensing
5410 Grosvenor Lane, Suite 210, Bethesda, MD 20814-2160, USA.

Battelle
Pacific Northwest Laboratories, Battelle Boulevard, PO Box 999, Richland, WA, 99352, USA.

Cambridge Committee for Aerial Photography (CUCAP)
Mond Building, Free School Lane, Cambridge CB2 3RF, UK.

Canada Centre for Remote Sensing
588 Booth Street, Ottawa, Ontario, K1A 0Y7, Canada.

CEN
100 Franklin Square Drive STE210, Somerset, NJ 08873, USA.

Department of Archaeology
National University of Ireland, Cork, Co. Cork, Republic of Ireland.

Department of Computer Science
National University of Ireland, Maynooth, Co. Kildare, Republic of Ireland.

Department of Geography
National University of Ireland, Maynooth, Co. Kildare, Republic of Ireland.

Discovery Programme
13 Hatch Street Lower, Dublin 2, Republic of Ireland.

ERA-Maptec
36 Dame St, Dublin 2, Republic of Ireland.

ESA/ESRIN
Via Galileo Galilei, I-00044 Frascati, Italy.

Goddard Spaceflight Centre
Greenbelt, MD 20771, USA.

Goldin–Rudahl Systems Inc.
6 University Drive, Amherst, MA 01002, USA.

Hunting Aerofilms Limited and Hunting Technical Services
Gate Studios, Station Road, Borehamwood, Hertfordshire WD6 1EJ, UK.

IS Limited
Atlas House, Atlas Business Centre, Simonsway, Manchester M22 5HF, UK.

Jet Propulsion Laboratory
California Institute of Technology, 4800 Oak Grove Drive, Pasadena, CA 91109-8099, USA.

Johnston Spaceflight Center
Houston, TX 77058, USA.

National Aeronautics and Space Administration
Headquarters, Washington DC 20546-0001, USA.

National Oceanic and Atmospheric Administration
Satellite Data Services, World Weather Building, Washington DC 20230, USA.

National Remote Sensing Centre
Delta House, Southwood Crescent, Southwood, Farnborough, Hampshire GU14 0NL, UK.

National Snow and Ice Data Center
CIRES CB 449, University of Colorado, Boulder, CO 80309-0449, USA.

National Space Development Agency of Japan
2-4-1 Hamamatsu-Cho, Minato-Ku, Tokyo, Japan.

Natural Resources Institute
University of Greenwich, Medway Campus, Chatham Maritime, Kent ME4 4AW, UK.

Ordnance Survey of Ireland
Phoenix Park, Dublin 6, Republic of Ireland.

Ordnance Survey of Northern Ireland
Colby House, Stranmillis Court, Belfast BT9 5BJ, Northern Ireland.

RADARSAT International Inc.
Satellite Data Distribution Centre, 3851 Shell Road, Suite 200, Richmond, BC V6X 2W2, Canada.

Remote Sensing Unit
Department of Geography, University of Bristol, University Road, Bristol BS8 1SS, UK.

Sandia National Laboratories
PO Box 5800, Albuquerque, NM 87185-0529, USA.

School of Earth and Environmental Sciences
University of Greenwich, Medway Campus, Chatham Maritime, Kent ME4 4AW, UK.

Space Imaging
9361 Grant Street, Suite 500, Thornton, CO 80229, USA.

SPOT IMAGE
5 rue des Satellites, BP 4369, F-31030 Toulouse Cédex 4, France.
Tel. no. (33) 562194101.
Fax no. (33) 562194054.

Ten-to-Ten (DIGITECH) Ltd
Unit 6D, Aberystwyth Science Park, Aberystwyth SY23 3AH, Wales.

United States Geological Survey
EROS Data Center, Sioux Falls, SD 57198, USA.

APPENDIX F

FURTHER READING

Arino, O. and Melinotte, J. M. (1995) 'Fire Index Atlas', *Earth Observation Quarterly* 50: 11–16.

Avery, T. E. and Berlin, G. L. (1992) *Fundamentals of Remote Sensing and Airphoto Interpretation*, 5th edition, Englewood Cliffs, NJ: Prentice-Hall.

Barrett, E. C. and Curtis, L. F. (1992) *Introduction to Environmental Remote Sensing,* London: Chapman and Hall.

Bastin, G. N., Pickup, G. and Pearce, G. (1995) 'Utility of AVHRR data for land degradation assessment: a case study', *International Journal of Remote Sensing* 16, 4: 651–72.

Belward, A. S. and Kennedy P. J. (1994) 'The limitations and potential of AVHRR GAC data for continental scale fire studies', *International Journal of Remote Sensing* 15, 11: 2215–34.

Billings, A. (1993) *Optics, Optoelectronics and Photonics: engineering principles and applications*, New York: Prentice-Hall.

Bond, P. (1993) *Reaching for the Stars: the illustrated history of manned spaceflight*, London: Cassell.

Burnside, C. D. (1985) *Mapping from Aerial Photographs*, London: William Collins and Co. Ltd.

Calder, N. (1991) *Spaceship Earth*, London: Viking.

Campbell, J. B. (1996) *Introduction to Remote Sensing,* 2nd edition, London: Taylor and Francis.

Christensen, F. T., Lu, Q. M. and Pedersen, L. T. (1994) 'Multi-year sea ice mapping by thermal infrared radiometry', *International Journal of Remote Sensing* 15, 6: 1229–50.

Cimino, J. B., Holt, B. and Richardson, A. H. (1988) *The Shuttle Imaging Radar B (SIR-B) Experiment Report*, Pasadena: NASA Jet Propulsion Laboratory Publication 88-2.

Clark, P. (1988) *The Soviet Manned Space Programme*, London: Salamander Books.

Clarke, R. N., Swayze, G. A., Gallagher, A. J., King, T. V. V. and Calvin, W. N. (1993) *The U.S. Geological Survey, Digital Spectral Library: version 1: 0.2 to 3.0 microns*, US Geological Survey Open File Report 93-592.

Curran, P. (1985) *Principles of Remote Sensing,* Harlow: Longman.

Curran, P. C. (1994) 'Imaging spectrometry', *Progress in Physical Geography* 18, 2: 247–66.

Danson, F. M. and Plummer, S. E. (eds) (1995) *Advances in Environmental Remote Sensing*, Chichester: John Wiley and Sons Ltd.

Davies, M. E. and Murray, B. C. (1971) *The View from Space*, New York: Columbia University Press.

Dean, K. G., Engle, K., Lu, Z., Eichelberger, J., Neal, T. and Doukas, M. (1996) 'Use of SAR data to study active volcanoes in Alaska', *Earth Observation Quarterly* 53: 21–3.

Department of Energy (1995) *Arms Control and Non-Proliferation Technologies*, Livermore, CA: DOE/NN/ACNT-95B.

Dickinson, G. C. (1969) *Maps and Air Photographs*, London: Edward Arnold Ltd.

Drury, S. A. (1993) *Image Interpretation in Geology*, London: Chapman and Hall.

Drury, S. A. (1998) *Images of the Earth: a guide to remote sensing*, Oxford: Oxford Science Publications.

Ehrlich, D., Estes, J. E. and Singh, A. (1994) 'Applications of NOAA-AVHRR 1 km data for environmental monitoring', *International Journal of Remote Sensing* 15, 1: 145–61.

Elachi, C. (1980) 'Spaceborne imaging radar: geologic and oceanographic applications', *Science* 209: 1073–82.

Elachi, C., Bicknell, T., Jordan, R. L. and Wu, C. (1985) 'Spaceborne synthetic aperture imaging radars: applications, techniques and technology', in H. Lee and G. Wade (eds) *Imaging Technology*, New York: Institute of Electrical and Electronic Engineers.

Elachi, C., Cimino, J. B. and Settle, M. (1986) 'Overview of the Shuttle Imaging Radar-B preliminary scientific results', *Science* 232: 1511–16.

European Space Agency (1993) *ERS User Handbook*, Noordwijk, Netherlands: European Space Agency.

European Space Agency (1994) *From ERS-1 to ERS-2: destination Earth*, Noordwijk, Netherlands: European Space Agency.

European Space Agency (1995) *New Views of the Earth – scientific achievements of ERS-1*, Noordwijk, Netherlands: European Space Agency.

Ford, J. P., Cimino, J. B. and Elachi, C. (1983) *Space Shuttle Columbia Views the World with Imaging Radar: the SIR-A experiment,* Pasadena: NASA Jet Propulsion Laboratory Publication 82-95.

Ford, J. P., Cimino, J. B., Holt, B. and Ruzek, M. R. (1986) *Shuttle Imaging Radar Views the Earth from Challenger: the SIR-B experiment*, Pasadena: NASA Jet Propulsion Laboratory Publication 86-10.

Furniss, T. (1986) *Manned Spaceflight*, London: Janes' Publishing Co.

Gutman, G. and Ignatov, A. (1995) 'Global land monitoring from AVHRR', *International Journal of Remote Sensing* 16, 13: 2301–9.

Haack, D. and Jampoler, S. (1995) 'Colour composite comparisons for agricultural assessments', *International Journal of Remote Sensing* 16 (9): 1589–98.

Hame, T., Salli, A., Andersson, K. and Lohi, A. (1996) 'Forest biomass estimation in northern Europe using NOAA AVHRR data', *Earth Observation Quarterly* 52: 19–23.

Harries, J. E. (1995) *Earthwatch: the climate from space*, Chichester: John Wiley and Sons in association with Praxis Publishing Ltd.

Harrison, E. F., Minnus, P., Barkstrom, B. R. and Gibson, G. G. (1993) 'Radiation budget at the top of the atmosphere', in R. J. Gurney, J. L. Foster and C. L. Parkinson (eds) *Atlas of Satellite Observations Related to Global Change*, New York: Cambridge University Press.

Harvey, J. G. (1976) *Atmosphere and Ocean*, Horsham: Artemis Press Ltd.

Houghton, J. T. (1986) *The Physics of Atmospheres*, New York: Cambridge University Press.

Houghton, J. (1996) 'Fifty years of technology development', *Weather* 51, 3: 163–6.

Hunt, G. R. and Ashley, R. P. (1979) 'Spectra of altered rocks in the visible and near infrared', *Economic Geology* 74: 1613–29.

Hunt, G. R. and Salisbury, J. W. (1970) 'Visible and near-infrared spectra of minerals and rocks: I silicate minerals', *Modern Geology* 1: 283–300.

Hunt, G. R. and Salisbury, J. W. (1976a) 'Visible and near-infrared spectra of minerals and rocks: XI, sedimentary rocks', *Modern Geology* 5: 211–17.

Hunt, G. R. and Salisbury, J. W. (1976b) 'Visible and near-infrared spectra of minerals and rocks: XII, metamorphic rocks', *Modern Geology* 5: 219–28.

Hunt, G. R., Salisbury, J. W. and Lenhoff, C. J. (1973) 'Visible and near-infrared spectra of minerals and rocks: VII, acidic igneous rocks', *Modern Geology* 5: 217–24.

Hyman, A. H. (1996) 'Information presentation for new sensors: a focus on selected sensors of the Earth Observation System (EOS)', *Progress in Physical Geography* 20, 2: 146–58.

Kilford, W. K. (1979) *Elementary Air Survey*, London: Pitman Publishing Limited.

Kondratyev, K. Y., Buznikov, A. N. and Pokrovsky, O. M. (1996) *Global Change and Remote Sensing*, Chichester: Praxis Publishing Ltd.

Kovaly, J. J. (1977) 'High resolution radar fundamentals (synthetic aperture and pulse compression)' in E. Brookner (ed.) *Radar Technology*, Dedham, MA: Artech House Inc.

Kyle T. G. (1993) *Atmospheric Transmission: emission and scattering*, New York: Pergamon Press.

Legg, C. (1989) 'The potential of meteorological satellite data for monitoring flooding disasters', Farnborough: National Remote Sensing Centre, NRSC SP(89) WP31.

Legg, C. (1991) 'An investigation of the Arabian Gulf oil slick using NOAA AVHRR imagery', Farnborough: National Remote Sensing Centre, NRSC SP(91) WP19.

Legg, C. A. (1994) *Remote Sensing and Geographic Information Systems*, Chichester: Praxis Publishing Ltd.

Lillesand, T. M. and Kiefer, R. W. (1994) *Remote Sensing and Image Interpretation*, New York: John Wiley and Sons.

Lo, C. P. (1986) *Applied Remote Sensing*, New York: Longman.

Lothian, G. F. (1975) *Optics and its Uses*, New York: Van Nostrand Reinhold.

Lowman, P. D. (1980) 'The evolution of geological space photography', in B. S. Siegal and A. R Gillespie (eds) *Remote Sensing in Geology*, New York: John Wiley and Sons Inc., pp. 91–115.

Lutgens, F. K. and Tarbuck, E. J. (1986) *The Atmosphere: an introduction to meteorology*, Englewood Cliffs, NJ: Prentice-Hall.

McCloy, K. R. (1995) *Resource Management Information Systems: process and practice*, London: Taylor and Francis.

Meteorological Office (1997) 'Nimrod: the leading edge of now-casting', Bracknell: Meteorological Office Factsheet 97/867.

Moffitt, F. H. and Mikhail, E. M. (1980) *Photogrammetry*, New York: Harper and Row Publications.

Murphy, R. and Wadge, G. (1994) 'The effects of vegetation on the ability to map soils using imaging spectrometer data', *International Journal of Remote Sensing* 15, 1: 63–86.

Newkirk, D. (1990) *Almanac of Soviet Manned Space Flight*, Houston, TX: Gulf Publishing Co.

Nicolson, I. (1982) *Sputnik to Space Shuttle*, London: Sidgwick and Jackson.

Njoku, E. G. (1985) 'Passive microwave remote sensing of the Earth from space – a review', in H. Lee and G. Wade (eds) *Imaging Technology*, New York: Institute of Electrical and Electronic Engineers.

Open Universiteit (1989) *Remote Sensing*, Heerlem: Open Universiteit.

Pavlakis, P., Sieber, A. and Alexandry, S. (1996) 'Monitoring oil-spill pollution in the Mediterranean with ERS SAR', *Earth Observation Quarterly* 52: 8–11.

Pedrotti, F. L. and Pedrotti, L. S. (1993) *Introduction to Optics*, Englewood Cliffs, NJ: Prentice-Hall.

Pinker, R. and Karnieh, A. (1995) 'Characteristic spectral reflectance of a semi-arid environment', *International Journal of Remote Sensing* 16, 17: 1341–63.

Piwowar, J. M. and LeDrew, E. F. (1995) 'Hypertemporal analysis of remotely sensed sea-ice data for climate change studies', *Progress in Physical Geography* 19, 2: 216–42.

Porter, R. W. (1977) *The Versatile Satellite*, Oxford: Oxford University Press.

Price, J. (1995) 'Example of high resolution visible to near IR reflectance spectra and a standardized collation for remote sensing studies', *International Journal of Remote Sensing* 16, 6: 993–1000.

Richie, W., Wood, M., Wright, R. and Tait, D. (1988) *Surveying and Mapping for Field Scientists*, Harlow: Longman Group UK Ltd.

Rivard, B. and Arvidson, R. E. (1992) 'Utility of imaging spectrometry for lithologic mapping in Greenland', *Photogrammetric Engineering and Remote Sensing* 58, 7: 945–9.

Sabins, F. F. (1997) *Remote Sensing: principles and interpretation*, New York: W. H. Freeman and Co.

Shapland, D. and Rycroft, M. (1984) *Spacelab: research in space orbit*, Cambridge: Cambridge University Press.

Smith, S. D. (1995) *Optoelectronic Devices*, London: Prentice-Hall.

Stewart, J. B., Engman, E. T., Feddes, R. A. and Kerr, Y. (eds) (1996) *Scaling up in Hydrology Using Remote Sensing*, Chichester: John Wiley and Sons Ltd.

Twomey, S. (1977) *Atmospheric Aerosols*, New York: Elsevier Scientific Publishing Co.

Verger, F. (1994) 'Remote sensing of the Earth's resources', *Sistema Terra* 2, 56–62.

Verger, F. and Soubes, I. (1994) 'Evolution of remote sensing', *Sistema Terra* 3, 64–7.

Vincent, R. K. (1997) *Fundamentals of Geological and Environmental Remote Sensing*, Englewood Cliffs, NJ: Prentice-Hall.

Williams, J. C. C. (1969) *Simple Photogrammetry*, London: Academic Press.

Williams, J. (1995) *Geographic Information from Space*, Chichester: Praxis Publishing Ltd.

Wolf, P. R. (1983) *Elements of Photogrammetry with Air Photo Interpretation and Remote Sensing*, New York: McGraw-Hill Inc.

APPENDIX G

PREVIEW OF *INTRODUCTORY REMOTE SENSING: DIGITAL IMAGE PROCESSING AND APPLICATIONS*

Introductory Remote Sensing: Principals and Concepts has introduced the basic concepts of remote sensing and also includes a number of applications for which remote sensing is particularly well suited. However, in order to use the data obtained by the satellite to produce the images shown in this book, it was often necessary to process the data digitally. If this processing was not performed, then often no useful information could be obtained from the data. Indeed, in many instances it would not be possible even to view the imagery. The digital processing that needs to be performed in order to create images is considered in the companion volume and is an extremely important aspect of remote sensing.

A brief outline of remote sensing principles and the concept of a digital image is given in Chapter 1. Digital image processing can be considered under two broad headings: pre-processing the data and enhancing the data. Pre-processing is dealt with in Chapter 2 of the companion volume and considers:

Digital image histograms
Co-registration of data
Line-banding corrections
Line-dropout corrections
Geometric corrections
Atmospheric corrections
Solar illumination corrections
Digital data format.

Following an explanation of the pre-processing techniques, the various enhancement procedures that can be applied to the data are covered in Chapter 3. The procedures discussed include:

Contrast stretching
Density slicing
Ratio imagery
Filtering techniques
IHS transforms
Principal components analysis
Synergistic display and non-image spatial data
Change detection imagery
Classification procedures.

In addition, the hardware and software used in digital image processing are also discussed in this chapter.

More information is provided in Chapter 4 on environmental monitoring techniques that use remotely sensed data. These include rainfall modelling and the various types of vegetation indices that can be produced.

Chapter 5 covers a number of case studies in some detail. Topics include:

Rainfall monitoring in west Africa
Sea-level rise modelling
Tropical cyclone tracking
Detection of peat soils
Synergistic display of geophysical datasets for geological mapping.

Hands-on computer processing of remotely sensed data is extremely important in remote sensing. Included with the hardback companion volume is a modified version of DRAGON image processing software which will enable you:

to display, manipulate, measure and analyse remotely sensed images in single-band format and as three-band combinations in false colour composites;

to perform different types of contrast stretching on the imagery, including user-defined stretches;

to zoom in to look at the images in more detail and to find out the digital values associated with specific positions on the images;

to produce DN histograms that summarise the distribution of the digital values in one band;

to produce scatterplots of two bands which provide a qualitative view of the correlation between bands;

to measure linear and areal features on the images;

to produce ratio images;

to apply filters to the imagery including user-defined ones;

to perform unsupervised and supervised classification;

to save the results of any processing to a separate file.

Seventy-seven datasets are included with the companion volume, along with a number of explanatory practicals which will allow you:

to acquire first-hand experience of processing the data;

to produce many more images than can be included in the book;

to obtain a fuller understanding of the various enhancement procedures;

to incorporate digital image processing within a remote sensing course.

The appendices in the companion volume include:

Appendix A: Answers to Self-Assessment Tests
Appendix B: Acronyms Used in Remote Sensing
Appendix C: Glossary of Remote Sensing Terms
Appendix D: Remote Sensing Sources of information
Appendix E: Contributors to *Introductory Remote Sensing*
Appendix F: Further Reading

Practical Manual for Digital Image Processing

INDEX

Note: Figures and Tables are indicated (in this index) by *italic page numbers* and Boxes by **bold page numbers**; "Pl." means "Colour Plate..." and "RS" = "remote sensing"

SPECIAL OFFER FOR READERS OF
INTRODUCTORY REMOTE SENSING⃰

You can get a copy of the full DRAGON/ips® Academic Edition remote sensing software system for a fraction of the normal price.

DRAGON Academic Edition provides image processing capabilities far beyond what you have seen in the restricted version of DRAGON included with this book.

Features include:

- Larger image capacity (1,024 × 1,024)
- Full color, multi-image display and annotation
- Image import, export, capture and printing
- Supervised and unsupervised classification (clustering, maximum likelihood, etc.)
- Classification accuracy assessment
- Geometric correction and image-to-image registration
- On screen vector capture
- Principal components analysis
- GIS integration functions
- Both Microsoft Windows™ and MS DOS versions
        and much, much more!   (See the CD-ROM for detailed technical information.)

All this for only: US$ 350 (Windows and MS-DOS)
US$ 250 (MS-DOS only)

This represents more than a 60% discount off usual prices!

To take advantage of this offer, copy the coupon on the reverse, or print the order form from the CD-ROM, and send with payment to:

Goldlin-Rudahl Systems, Inc.
PMB 213, 6 University Drive, Suite 206
Amherst, MA 01002
U.S.A.

FAX: +1-413-549-6401
email: info@goldin-rudahl.com
WWW: http://www.goldin-rudahl.com

In the U.K., you may contact:

I.S. Ltd.
Atlas House, Atlas Business Centre
Simonsway
Manchester M22 5HF

Phone: 0161-499-7609
FAX: 0161-436-6690
email: isman@compuserve.com

⃰ Limited time offer: Discount prices guaranteed until January 1, 2002

Instructors: Please contact Goldin-Rudahl Systems for information on discounts on quantity purchases, the DRAGON Professional Suite, site licences, and other products.

Order Form

Please send me:

- [ ] DRAGON/ips® Academic Edition for Windows and MS-DOS (US$ 350)*
- [ ] DRAGON/ips® Academic Edition for MS-DOS (US$ 250)
  (Prices include shipping by air post.)

Payment type:

- [ ] Cheque payable by a U.S. Bank
- [ ] Visa
- [ ] Mastercard

Credit card number: _____ Expires: _____

Authorized signature: _____

Ship to:

Name: _____

Address: _____

_____

City: _____ State/Province/County: _____

Postal code: _____ Country: _____

Email: _____

Fax: _____

Mail to:

Goldin-Rudahl Systems, Inc.
PMB 213, 6 University Drive, Suite 206, Amherst, MA 01002 USA

Fax to:

(Credit card holders only)
Goldin-Rudahl Systems, Inc.
+1-413-549-6401

* Windows version requires at least 32,000-color gaphics (15 bit color) 32 MBytes memory and CD-ROM drive.
  MS-DOS version requires at least VGA graphics and 2 MBytes memory.